MERCURY CONTAMINATION:

A HUMAN TRAGEDY

PATRICIA A. D'ITRI

Associate Professor of
American Thought and Language
Michigan State University
East Lansing, Michigan

FRANK M. D'ITRI

Associate Professor of
Water Chemistry
Institute of Water Research
Michigan State University
East Lansing, Michigan

A WILEY-INTERSCIENCE PUBLICATION

JOHN WILEY & SONS
NEW YORK • LONDON • SIDNEY • TORONTO

Copyright © 1977 by John Wiley & Sons, Inc.

All rights reserved. Published simultaneously in Canada.

No part of this book may be reproduced by any means, nor transmitted, nor translated into a machine language without the written permission of the publisher.

Library of Congress Cataloging in Publication Data

D'Itri, Patricia A
    Mercury contamination.

    (Environmental science and technology)
    "A Wiley-Interscience publication."
    Includes bibliographical references and index.
    1. Mercury—Environmental aspects. I. D'Itri, Frank M., joint author. II. Title.

TD196.M38D57   363.6   76-58478
ISBN 0-471-02654-9

Printed in the United States of America

10 9 8 7 6 5 4 3 2 1   78-5380

MERCURY CONTAMINATION

FLYING MERCURY

QUICKSILVER

Can you hear the tortured screams yet?
Do you see the twisted limbs?
Does it frighten you completely?
Then you're ready to begin.

A sideshow aftermath
from eating pink-dyed grain
or fleshy, beaded greyshine fish
from Sacred Mother waters.

The numbers grow
by family:
Minamata mothers, Kenora fathers,
Alamogordo's son,
Dark-eyed Iraqian daughters.

Do the victims' acrobatic
poses merit your surprise?
Then avoid the daily papers
and avert your T.V. eyes.
Forget the dark warnings
until your tragedy arrives.

—By Judith Ecker

1

2

The Chisso Company (1) released methylmercury into Minamata Bay where it concentrated in fish that were consumed by local fishermen and their families (2). Families of poor fishermen became crippled and died from eating methylmercury-con-

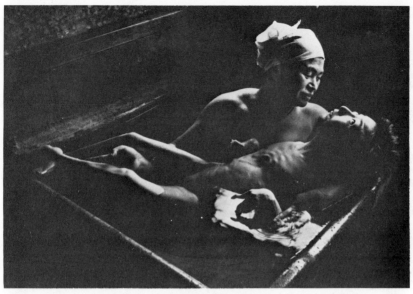

3

4

taminated fish (3). Then the victims rebelled and demanded compensation from the Chisso Company (4).

5

6

7

Barney Lamm and three former guides return to the deserted Ball Lake Lodge in 1976 (5), while American tourists continue to fish and consume their catches, even on the highly contaminated Clay Lake (6), and while Indians guide at other lodges and eat fish daily at shore lunches (7). As Indian children are born with crippling illnesses that resemble the symptoms of Minamata disease (8), Minamata disease

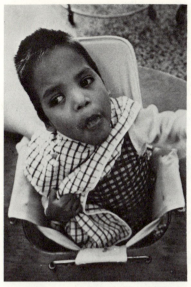

8

PHOTO BY HIRO MIYAMATSU

9

PHOTO BY KOICHI ENISHI

10

PHOTO BY AILEEN M. SMITH

victims come to Toronto, Ontario, to join their protests against government inactivity (9), and young Japanese and Indian victims pledge their assistance to one another in front of the factory that contaminated hundreds of miles of the Wabigoon–English River system in Ontario, Canada (10).

11

PHOTO BY KOICHI ENISHI

12

PHOTO BY HIRO MIYAMATSU

Scientists calculate that fish will remain contaminated for decades on the Wabigoon–English River chain that remains the primary natural source of food for the Indians who live at White Dog (11) and Grassy Narrows Reserves (12) in Ontario, Canada. Poverty dictates that mothers must continue to prepare and eat foods that cripple and retard their unborn babies as well as the rest of the family, as when mercury-

13

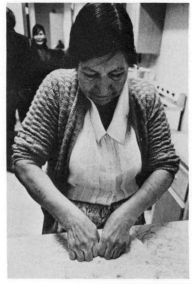
14

PHOTO BY HIRO MIYAMATSU

15

contaminated wheat was made into bread in Iraq (13), and Indian women added the eggs of contaminated fish to bread for their families (14). Thus like the primitive millstone used to grind the lethal grain into flour in Iraq (15), mercury-contaminated food remains a slow but relentless threat to people who must avail themselves of the traditional fish and loaves.

This book is dedicated
to the victims of Minamata disease
and to the D'Itri children:
Michael, Angella, Patricia, and Julie.

SERIES PREFACE

Environmental Science and Technology

The Environmental Science and Technology Series of Monographs, Text-books, and Advances is devoted to the study of the quality of the environment and to the technology of its conservation. Environmental science therefore relates to the chemical, physical, and biological changes in the environment through contamination or modification, to the physical nature and biological behavior of air, water, soil, food, and waste as they are affected by man's agricultural, industrial, and social activities, and to the application of science and technology to the control and improvement of environmental quality.

The deterioration of environmental quality, which began when man first collected into villages and utilized fire, has existed as a serious problem under the ever-increasing impacts of exponentially increasing population and of industrializing society. Environmental contamination of air, water, soil, and food has become a threat to the continued existence of many plant and animal communities of the ecosystem and may ultimately threaten the very survival of the human race.

It seems clear that if we are to preserve for future generations some semblance of the biological order of the world of the past and hope to improve on the deteriorating standards of urban public health, environmental science and technology must quickly come to play a dominant role in designing our social and industrial structure for tomorrow. Scientifically rigorous criteria of environmental quality must be developed. Based in part on these criteria, realistic standards must be established and our technological progress must be tailored to meet them. It is obvious that civilization will continue to require increasing amounts of fuel, transportation, industrial chemicals, fertilizers, pesticides, and countless other products; and that it will continue to produce waste products of all descriptions. What is urgently needed is a total systems approach to modern civilization through which the pooled talents of scientists and engineers, in cooperation with social scientists and the medical profession, can be focused on the development of order and equilibrium in the presently disparate segments of the human environment. Most of the skills and tools that are needed are already in existence. We surely have a right to hope a technology that has created such manifold environmental problems is also capable of solving them. It is our hope that this Series in Environmental Sciences and Technology will not only serve to make this challenge more explicit to the established professionals, but that it also will help to stimulate the student toward the career opportunities in this vital area.

Robert L. Metcalf
James N. Pitts, Jr.
Werner Stumm

PREFACE

Knowledge will forever govern ignorance and a people who mean to be their own governors must arm themselves with the power knowledge gives. A popular government without popular information or the means of acquiring it, is but a prologue to a farce or a tragedy or perhaps both.

—*James Madison*

After the cats began to dance, first poverty-stricken Japanese fishermen and then Canadian Indians saw their children born retarded and crippled while their own brain cells and central nervous systems were destroyed because of methylmercury poisoning from eating contaminated fish. This book traces the history of these and other human tragedies caused by alkylmercurials directly or as a result of environmental contamination as well as the applications of the liquid element in industry and medicine. Although some of these processes are now obsolete, others have continued over the centuries, sometimes with very little assessment of their impact on people or the environment. In addition to the three sections of the book on environmental contamination, industrial uses, and medical applications, the fourth section summarizes medical treatments, some methods of removing waste mercury from the environment, legislation, and the political pattern that has emerged in a number of mercury-pollution epidemics. Jun Ui's afterword is a thoughtful assessment of the problem. As one of the mos courageous and outspoken scientists of our time, he has battled the bureaucracies of several countries on behalf of the growing ranks of Minamata disease victims; and the authors' debt to him is profound.

As a case study of present and previous applications of one still virtually indispensable element, this book is intended for readers who want to be better informed about the widely publicized mercury contamination problem of recent years, its historical background in industry and medicine, and the consequent activities related to this contaminant in a number of other countries as well as the United States. The book was not written for scientists who conduct research into mercury pollution or for industrialists and politicians who must deal with the problems that have been caused. Rather, it should be of interest to citizens who are concerned about the degradation of the environment, and it could serve as a textbook for advanced high school and college classes to explain the uses and abuses of this element in itself and as a prototype of other pollutants. Besides conveying information, it is hoped that this book will help overcome apathy and complacency so that we will all act to exert the necessary pressure to insist that industry and government halt further contamination and human suffering, to prevent those already stricken from serving as warning organisms for the rest of us if this kind of contamination is allowed to continue.

The authors wish to acknowledge assistance received from their respective departments at Michigan State University as well as numerous individuals and other organizations. Within the university we are most grateful for assistance from The Institute of Water Research, the Department of American Thought and Language, the Department of Fisheries and Wildlife, and James Madison College. In addition, we appreciate the assistance rendered by Jun Ui, Barney and Marion Lamm, Tom and Judith Ecker, the Rockefeller Foundation, the Canadian National Indian Brotherhood, and the Ontario Public Interest Research Group (OPIRG), University of Warterloo, Ontario.

Of course, the authors alone are responsible for any errors included herein. They conclude with the hope that the readers share the commitment to insist that industry, government, and scientific research be conducted with human betterment as the goal rather than as ends in themselves. This, finally, is what *Mercury Pollution: A Human Tragedy* is about.

PATRICIA A. D'ITRI
FRANK M. D'ITRI

East Lansing, Michigan
November 1976

CONTENTS

CONVERSION UNITS

UNITS OF WEIGHT

1 flask	$= 76$ lb
1 kilogram (kg)	$= 1000$ g $= 2.205$ lb
1 gram (g)	$= 10^{-3}$ kg $= 0.035$ oz
1 milligram (mg)	$= 10^{-3}$ g
1 microgram (μg)	$= 10^{-6}$ g
1 nanogram (ng)	$= 10^{-9}$ g
1 picogram (pg)	$= 10^{-12}$ g

UNITS OF VOLUME

1 liter (1)	$= 1.056$ qt
1 liter (1)	$= 1000$ ml
1 milliliter (ml)	$= 10^{-3}$ 1
1 microliter (μl)	$= 10^{-6}$ 1
1 cubic meter (m^3)	$= 1000$ 1

UNITS OF CONCENTRATION

Weight-to-Weight Relationships
(For plants, foods, flesh, and all solids in general)

1 part per million (ppm) $= 1$ mg/kg or 1 μg/g
1 part per billion (ppb) $= 1$ μg/kg or 1 ng/g
1 part per trillion (ppt) $= 1$ ng/kg or 1 pg/g

Weight-to-Volume Relationships

(Usually for water or other liquids in general)
1 mg/liter = approximately 1 ppm on weight-to-weight basis
1 μg/liter = approximately 1 ppb on weight-to-weight basis
1 ng/liter = approximately 1 ppt on weight-to-weight basis

Weight-to-Volume Relationships

(Usually for air or other gases in general)

1 μg/m^3 = approximately 0.8 ppb on weight-to-weight basis
1 ng/m^3 = approximately 0.8 ppt on weight-to-weight basis

MERCURY CONTAMINATION

INTRODUCTION

OVERVIEW

In the final analysis, there are no non-hazardous substances; there are only non-hazardous ways to use substances, or levels of substances whose use poses no hazard. However, we are not presently faced with widespread, serious human health hazards from these substances. Our concern today is primarily about future generations—that we do not, by our short sightedness, today, condemn future generations to irreversible hazardous health effects (1).

The future had come to Japanese fishermen at Minimata Bay nearly 20 years before Surgeon General Steinfeld made these remarks to the Hart Subcommittee hearings in 1970 when mercury contamination was thought to be a local problem in Lake St. Clair instead of a national environmental issue. Even after the Huckleby family was poisoned at Alamogordo, New Mexico, Americans had not really been alerted that mercury and other environmental contaminants pose a more insidious hazard than the atomic bomb exploded near that small community 25 years earlier. Although the Big Bang was initially more frightening, the toxic substances that have accumulated in the environment, especially in the post-World War II era of industrial expansion, could ultimately be more difficult to control and remove.

The deaths of Japanese fishermen at Minamata first alerted scientists that toxic quantities of methylmercury could concentrate in fish. From 50 to 30,000 times more methylmercury was found in their bodies than in the waterways in which they lived. Yet many of the fish looked perfectly healthy. Human beings, in turn, concentrated the compound as it was passed upward from the lower echelons of the food chain. Whether the methylmercury was consumed indirectly in fish or other animals or directly by eating methylmercury-treated seed grain, the cause of a major epidemic in Iraq, the results were the same: permanent damage to the brain and central nervous system.

The afflicted person has no symptoms during a latent period in which the deadly compound gradually concentrates. By the time the injury is noticeable, the damage has already been done. When the exposure is constant, such as eating highly contaminated fish daily for as little as three months, the person can become crippled and blind or even die of acute poisoning. The early symptoms may be minor emotional disorders such as a tendency toward depression, excitability, headaches, and fatigue. These are followed by decreasing physical coordination and loss of memory. Pregnant women are often spared from poisoning, because the methylmercury quickly crosses the placental barrier to accumulate in their unborn children instead.

With less frequent or chronic exposure to a much lower dose of the contaminants, the same symptoms appear more gradually. The methylmercury concentrates in the brain and remains there for a very long time. Even a 2-week fishing trip will leave the human blood mercury levels higher after two years if fish with up to 10 ppm of mercury are consumed. When the brain cells are gradually destroyed, senility is advanced because the reserve cells are needed to replace them. At Minamata people with no other symptoms of disease are gradually showing the signs of aging in their middle years that are normally seen in their elders. With this gradual intoxication methylmercury poisoning is sometimes only confirmed after the individuals die. Among those recognizable symptoms, hundreds of cases have been confirmed around Minamata Bay, and thousands wait to be recognized and receive compensation. They range from congenitally retarded children to family members left without support by the death of their father from acute poisoning. In many instances several members of the same family have been stricken.

As mercury concentrates and translocates up the food chain to poison either directly or indirectly, Swedish scientists have discovered that elemental mercury and inorganic salts or compounds also can be degraded and methylated either chemically or biologically in the waterways or in the bodies of fish and animals. Whereas at Minamata methylmercury was an unintentional byproduct of the industrial process, the effect on the food chain can be the same when large quantities of elemental mercury have been discharged from industries, notably mercury cell chloralkali plants. Thousands of tons of the element were discharged in their waste effluent and settled on the bottoms of the waterways. In that setting even the bacteria are influenced by the company they keep; they absorb and methylate the mercury before returning it to the water to concentrate in fish.

When fish are no longer safe for human consumption, restrictions on commercial and sport fishing have had a severe economic impact on local areas that depend on tourism or the commercial fishing industry. This problem has now arisen in the United States and Canada as well as Japan and Sweden. In each case the most tragic victims are people who are too poor to purchase other food and continue to eat fish from the local waters, notably the poor fishermen at Minamata and the Indians in Ontario, Canada.

By far the most widespread and serious mercury problem is posed by industries that utilize and discharge enormous quantities into the air or water. Some major industries like felting hats have declined, but the market for mercury to manufacture electrical and mechanical equipment as well as chlorine and caustic soda has grown steadily. Moreover, some ancient industrial processes also continue with updated technology. Among these are mining the element itself and smelting other ores.

The workers are exposed to inorganic mercury poisoning, the traditional disease associated with the element, its salts, and especially its vapors. The symptoms are tremors, loose teeth, minor psychological changes, headaches, and fatigue. But the risks can be limited by rotating miners to other jobs while a portion of the mercury is excreted from their bodies. Moreover, vapor levels in the air can be monitored, and the workers' blood and urine mercury levels can be checked to detect poisoning symptoms early. Because prevention is simpler and the symptoms are reversible, inorganic mercury is much less of a threat than the methylmercury that continues to concentrate in the human body and eventually destroys the brain and central nervous system. However, as small doses of inorganic mercury are absorbed in thousands of common uses, the body may convert some of it into the more hazardous organic form to remain as part of the total body burden.

Mercury medications are a minor but long-standing source of exposure. They cause sensitivity reactions in some individuals and add to the total mercury in all. However, the long medical history of mercury is most interesting, because it demonstrates a widespread acceptance of the element's curative powers based on tradition rather than fact. From the first mercury ointments applied in ancient Greece, misconceptions about its curative powers have caused suffering among hundreds of thousands of individuals over many centuries. Generations of physicians reinforced the belief that mercury was an effective ingredient in ointments to treat skin irritations and to cure syphilis. Since the 1940s penicillin has replaced mercury to treat social diseases, but the public continues to purchase mercurial antiseptics and ointments despite more effective replacements. However, many medical compounds containing inorganic and organic mercury are being discontinued.

In a way the disaster at Minamata initiated a kind of reverse science. Researchers had to look at what had already happened to wastes discharged into the environment to assess the potential damage and to determine what should be done about it. Even to identify the original condition of the environment, much less to restore it, might be impossible, because the contamination is now so extensive. Vast quantities of mercury have been removed from a few natural repositories, funneled through manmade enterprises, and distributed in the environment around the most densely populated areas. However, such problems cannot be isolated geographically or socially, because modern industrial technology has also contaminated remote rural areas. Overcoming this threat to human health and the environment, then, will require the development of new reclamation and preventive techniques to restore and preserve the environment. To accomplish this, old patterns of carelessness and ignorance must change.

Historically, mercury has been widely applied in industry, agriculture, and medicine for centuries in which progress has generally been advanced through sketchy trial-and-error methods followed by the discovery of their accompanying negative side effects. New technical applications for mercury evolved first, followed by related evidence of the less-advantageous side effects. Thus the history of mercury is a prototype of the way many elements and compounds have frequently been applied and discarded with little regard for the consequences to human health and the environment.

We now know they are linked by complex cause-and-effect relationships between overlapping technological applications and their negative ramifications as the ecosystem has been restructured through human intervention. This overview of mercury's role in the technological evolution of modern society begins with a brief account of its transition from a knavish Roman god to a mundane but useful element, essentially the transfer from the priorities of religion to those of science and the consequent applications of mercury in industry.

HISTORY

From the ancient god to the modern spaceship of the same name, the liquid element mercury has occupied an unusual niche in human history. With over 3000 applications in medicine and industry, the element's status has declined as its utility increased. The name Mercury originally identified the Roman god of commerce and the market, protector of traders. The boyish figure with wings on his helmet and sandals is more familiar than Hermes, his predecessor in Greek mythology. Mercury was one of the most popular Roman gods and appears in more mythological tales than any other. Perhaps his popularity stemmed from the fact that he was the shrewdest and most cunning of the gods; he was a master thief who started his knavery before he was a day old by stealing Apollo's herds. His grace and swiftness earned him a place as Jupiter's messenger, but Mercury was also the god of science and the patron of travelers as well as rogues, vagabonds, and thieves (2). Shakespeare recalled his dual roles as messenger and thief.

> Delay leads impotent and snail-pac'd beggary.
> Then fiery expedition be my wing.
> Jove's Mercury, and herald for a king!
>
> *Richard the Third*, IV, iii, 53–55

> My father nam'd me Autolycus, who being,
> as I am, litter'd under Mercury, was likewise
> a snapper-up of unconsidered trifles.
>
> *The Winter's Tale*, IV, iii, 24–26

An additional role that the Romans assigned to their mythological favorite was to act as the Divine Herald who guided dead souls to their final rest (2), a mission that Mercury's namesake element has unfortunately also sometimes performed when cast in the many molds dictated by the marketplace. Thus it is ironic and appropriate that quicksilver has retained the name of the mischievous Roman god.

The most ancient specimen of quicksilver was found in a little ceremonial cup in an Egyptian tomb at Kurna. It dates from the fifteenth or sixteenth century B.C. (3). However, cinnabar ore, the primary mercury-bearing ore composed of mercuric sulfide, was made into red paint long before the process of refining it into mercury was discovered. Precolonial American Indians dug the red ore from caves in the mountains of California and South America to make the war paint with which they decorated their faces for tribal ceremonies.

In ancient Greece a debate long raged over whether mercury was refined from the ore of cinnabaris or minium. In his 37-volume *Historic Naturalis*, Caius Plinius Secundus (born 23? A.D., died 79 A.D.) contended that the two had been confused and that cinnabaris was really the thick matter which issues from the dragon when crushed beneath the weight of the dying elephant, mixed with the blood of either animal (4). He concluded that the dragon's blood mixture was a useful medication but that minium was the ore mined in Spain to obtain quicksilver (5). Subsequently, both Theophrastus and Dioscorides recalled that earlier Greeks had believed that the red paint made from cinnabar ore was the blood of dragons. When the dispute was finally settled many years later, cinnabar was identified as mercury ore and minium as red oxide of lead, whereas dragon's blood had passed completely out of favor.

Despite disputes over its origin, the Greeks knew how to recover mercury from cinnabar ore at least by the third century B.C. Pliny the Younger offered a method of purifying the element by squeezing it through leather, whereas Theophrastus of Lesbos described the more widely practiced method of heating earthern vessels containing cinnabar until the vapors were driven off to be cooled and condensed into elemental mercury. Nonetheless, Galenus (210 to circa 131 B.C.) later acknowledged that practically nothing was known about the heavy metal. Aristotle described only limited applications in medications and religious ceremonies. The modern scientific symbol Hg is derived from the Greek name *Hydrargyrum*, which means liquid silver. The Romans similarly named it *Argentum*

Vivum, Latin for "live" or "quick" silver, because mercury was the only element then known to be liquid at ordinary room temperatures.

The Romans extended the Greek list of mercury applications, particularly after the Punic Wars when large quantities of cinnabar were transported to Rome. The ore was often reduced to a powder to make the highly prized pigment vermilion. Roman villas were decorated with this red coloring that was also included in rouges and other beauty products (6). Vermilion continues to be regarded as a high-grade paint pigment in the twentieth century. The Romans required more mercury after they learned about amalgamation. Vitruvius, the architect, described how fabrics were burned and the gold threads retrieved with mercury (7). By 77 A.D. 10,000 pounds of mercury were imported to Rome each year for this purpose. Most of it came from Spanish mines where Roman slaves and convicted criminals were sent to labor and consequently die by the thousands from exposure to mercury fumes (5).

In the Far East the knowledge of mercury was also gradually extended, although ancient historical accounts left by the Indians and Chinese are difficult to date because they were subsequently revised and updated (8). A relief map of China depicted oceans and rivers as liquid quicksilver a century after Aristotle lived (9). The early Chinese believed medications containing cinnabar and mercury could prolong life. Consequently, several emperors were reputed to have died from mercury poisoning in futile attempts to attain immortality (10).

By the Middle Ages western societies had rejected the ancient worshippers of the Roman god Mercury as heathens who put a human shape on a god "with wings to his feet." Then Christians attempted to reconcile contemporary scientific concepts with their religious principles. In the medieval cosmos the Earth was believed to be located at the center with the Moon above and then the planet Mercury in a geographical progression toward heaven. Consequently, astronomy was correlated with "the great mistery [sic] of resurrection." Whereas the Gallenists supported their belief in resurrection with St. Paul's argument to the Corinthians (1 Cor. 15:35), the chemists demonstrated the same concept with the teachings of Moses and the scientific process of mercury distillation. Since all terrestrial bodies were believed to consist of salt, sulfur, and mercury, they thought it reasonable that all terrestrial bodies could vaporize. As heat applied to cinnabar ore releases vapors that condense as purified mercury at the top of the vessel, at death human spirits were believed to vaporize and ascend to the Moon where the unpurified refuse would either remain in limbo, evaporate, or be annihilated in the final conflagration. Meanwhile, the purified spirits would ascend to the planet Mercury, an intermediate sphere, "til God (the Worlds great Chimist [sic]) thinks fit to dispose of them at the general Resurrection, or particular (as he thinks fit)" (11).

One medieval author, Lazarus Erckern, acknowledged that he could only guess whether Mercury itself was a hell, heaven, or limbo, but he was convinced that the spirits were appropriately categorized during their stay on the planet or in its surrounding orbit while they awaited God's will. Careful calculations in what he called "German miles" supported his contention that Mercury and its environs could easily contain all the spirits that could possibly ascend from Earth to await the Last Judgment. Although Mercury is only 442 German miles in diameter, the total orbit includes 62,999,698 German miles from the Moon below to the higher orbit of the planet Venus. Figure 1 depicts Erckern's conception of the medieval

Figure 1. Lazarus Erckern's view of the medieval universe. From *Spagyrik Laws* by L. Erckern, 1683.

cosmos. The Earth is at the center with the Moon and Mercury in the surrounding circles above it and below the outer ring. The darkened lower sphere of the Moon retains the refuse from earthly beings when their purified spirits ascend to the clear, upper sphere of Mercury to await the Last Judgment. Although Kepler reportedly disagreed with these computations on the heavenly spheres, Erckern remained convinced that the great spaces between planets must serve some Divine Purpose such as the one he outlines (11).

Besides explaining the spiritual allocation of space, Mercury's heavenly role was also justified by its long tradition. Since the symbol for the planet and the metal were the same and both had supposedly been revealed to the ancient Chaldeans and Egyptians, Erckern reasoned that the astrological characters of the seven planets originated at a time when more was known about their sympathies and concurring operations. God had supposedly revealed them to the early Hebrew authors who described seven planets, seven metals, and the figure of Mercury. Consequently, both the earlier and later symbols for Mercury were assumed to contain great mysteries as yet unrevealed to the faithful of the Middle Ages.

Thus in the transition from Roman mythology to Christianity, the god Mercury was replaced by the planet. As the chemical properties of the element were redefined, the sign of the planet Hermes was transferred from tin to mercury between the sixth and seventh centuries A.D. The symbol named for the Greek god was a circle with a dot in the center. In the Middle Ages it also designated the philosopher's stone, the sun, and gold (12). It signified that the liquid element was itself an incomplete but essential ingredient needed to convert base metals into gold. By the third century A.D., as the early science of alchemy evolved, a member of Cleopatra's court included mercury in an exotic process for making gold. This plan is reproduced in Figure 2.

The design formed of three concentric circles (A) has signs in the center for gold, silver (moon with appendage), and mercury (small moon). The Greek inscription in the inner circle reads "The serpent becomes one, the one with venom, after the two signs." The exterior circle is inscribed "This one is in himself complete; and if he is not in himself complete, he is not this one." The small figures are chemical apparatuses (B), and the large piece of equipment (C) at the lower right is a distiller set over a furnace. At the lower left the serpent Outroboros (D) encircles the words "This one is himself complete" (12–13).

Subsequently, the reknowned medieval alchemist, Paracelsus, reaffirmed

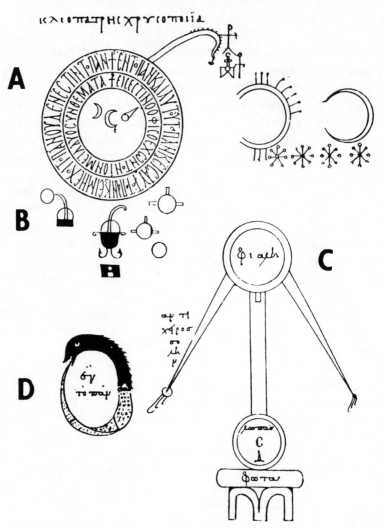

Figure 2. The gold-making art of Cleopatra's court. From *Chimie des anciens* by M. Berthelot, 1889.

10

the theory that mercury was an incomplete metal, an imperfect liquid that lacked the power to coagulate (13). Paracelsus concluded that all metals were liquid mercury up to the midpoint in their formation, but the proper application of fire would convert this element into other metals. Nonetheless, he was unable to transform mercury into gold despite hundreds of experiments (14), presumably because Paracelsus lacked the philosopher's stone that alchemists believed was the other magic ingredient required to complete the conversion.

Faith in mercury's magical powers was extended to medicine with new efforts to cure syphilis. Such notables as Napoleon, Ivan the Terrible, Robert Burns, and countless kings and noblemen are believed to have been shepherded to the nether world by mercurialists in vain attempts to cure the dread disease. Thus mercury was also heralded by poets to symbolize the end of love. Shakespeare observed that "The words of Mercury are harsh after the songs of Apollo" (*Love's Labor's Lost*, V, ii, 924–925), and the same theme appears in Alexander Pope's paraphrase of Chaucer's prologue to the Wife of Bath's tale.

> Love seldom haunts the breast where
>   learning lies,
> And Venus sets ere Mercury can rise.

Even in literature, however, mercury's status declined from threatening to absurd. In 1947 a fictional biography of the "Poet's Poetess," Sarah Binks, the sweet songstress of Saskatchewan, ends with Sarah's death caused by swallowing mercury from a thermometer designed to take the temperature of horses (15). Although myths and legends demoted mercury from the grandeur of the early gods to the gross parody of modern fiction, the sometimes bizarre and often careless disregard with which the element has been applied and discarded over the centuries has caused a progression of mischievous and often deadly turns of events befitting Jupiter's young messenger with the winged feet.

I

ENVIRONMENTAL CONTAMINATION

1

THE JAPANESE TRAGEDY

WHERE THE MODERN STORY BEGINS

The steadily mounting environmental contamination by mercury was ignored until a tragic series of events occurred in Japan. In the beginning many cats were seen to dance in the small fishing villages along Minamata Bay on Kyushu Island. They clearly were mad, because they screamed incessantly and often ended their dance and their lives by flinging themselves into the sea—indeed strange behavior for cats. This activity was first observed in 1953, and by 1960 the nervous tremors that preceded the dance were familiar not only in cats, but also in birds, fish, pigs, and dogs. Crows would fall out of the sky and flop about helplessly on the ground. Most of them died. But greater terror was aroused as human beings were also stricken, often several members of one family. Fearing that they might have a shameful infectious disease, the poor fishermen kept their tragedy to themselves for 3 years.

In 1956 Mrs. Watanabe brought her 6-year-old daughter, Matsuyio, to the pediatrics clinic of the hospital at the Shin Nihon Chisso Company, a large chemical factory at Minamata city. The doctors concluded that the child's nervous system had been afflicted, but they were at a loss to diagnose the cause of the mysterious disease that also struck Mrs. Watanabe's younger child, Eiishi, a week later. When four more patients were admitted to the municipal hospital, they were temporarily placed in an isolation ward because doctors suspected that they might have the contagious *Encephalitis japonica*, although they had no fever and their symptoms seemed to develop more slowly. Since all the victims lived near Minamata Bay, the syndrome was soon named Minamata disease after that small inlet (16) (see Figure 3).

When the medical staff at the Minamata Health Center examined physicians' records in the region, they learned that at least 30 other people suffered from the same symptoms. With a resident population calculated at 10,119 in the sea villages, the disease seemed to be reaching epidemic proportions. Consequently, a committee was quickly formed to find the cause; and four months later, on August 24, 1956, the medical school of Kumamoto University was commissioned to treat the patients and undertake a field study. Their work still continues.

An initial list of 52 patients was compiled among 40 families. From the first victim in December 1953 to October 1960, the prefectural health department recognized 121 patients: 68 adults, 30 children, and 23 fetal victims. Among the 46 deaths, one-half were adults, one-third were children, and 1 out of 8 were congenital cases. In a second epidemic at Niigata in 1964, 30 people were initially stricken and 5 died. By 1970 an additional 17 patients were added with one death and one congenital case (17). More subacute and chronic symptoms were identified in this second epidemic, just as the list lengthens at Minamata each year. Consequently, by the summer of 1973 over 850 Minamata disease victims had been identified, and the number is expected to be several times higher in the future, especially as more outbreaks are discovered (18). A third epidemic was reported at Ariake-cho in the Amakusa District on the Ariake Sea in May 1973 (18–19). Figure 3, a map of Japan, shows the locations of these contamination sites. Some evidence now indicates that other poisonings may have gone undiagnosed as early as 1946. And cats with symptoms have been observed in places distant from Minamata where no human illnesses have yet been diagnosed (20).

Since cats had similar symptoms, they were initially thought to have an infectious bacteria or virus that was being transmitted to the people. But the causative agent could also be another chemical, plant, or animal poison, bacterial toxin, or a food-borne infection. Thus the Kumamoto medical team were like detectives at the scene of a crime as they first interviewed residents of the small fishing villages. Then they compared the patients in 40 afflicted families with 68 control families through ques-

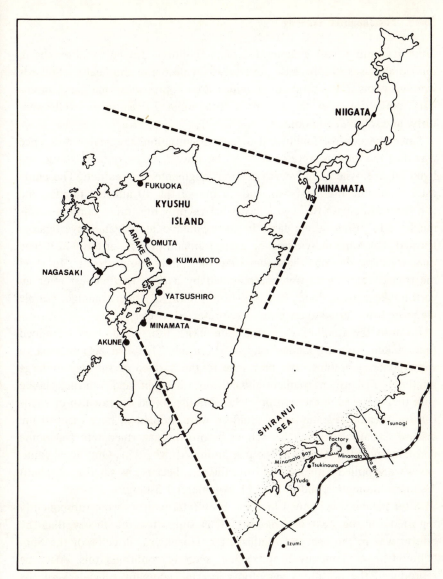

Figure 3. Kyushu Island and Minamata Bay, Japan.

17

tionnaires and a mass examination of the children. Living conditions were immediately suspect, because poverty was widespread and health standards were very low in the rural communities. The homes were small and unsanitary, and drinking water was often contaminated because the wells were dirty, uncovered, and polluted.

Since the control families also lacked sanitation, the major difference was that the victims were usually of the lowest socioeconomic status and lived on rice, wheat, and sweet potatoes augmented by seafood. The poorest families ate the most fish, and 25 out of the 40 afflicted families ate fish from Minamata Bay every day, whereas only four other families ate as much (21). Eight years later in the epidemic at Niigata, investigators learned that some individuals ate up to a pound of fish per day (22). The quantity depended on what other food the family could provide. In 1964 the tragedy at Niigata was compounded by a poor rice harvest after an earthquake caused floods in the district. With less rice available, people ate more fish. Carp and dace were the preferred species.

Because the families of victims ate more fish, researchers examined those taken from Minamata Bay more closely. The fish also seemed to show some symptoms, and they rose to the surface of the bay in large numbers. Fishermen frequently went out at night and returned home with their catches in the morning when families often ate raw fish or shellfish for breakfast. Later, some children were thought to have escaped the disease, because they left for school before the fishermen returned home and consequently ate less of the contaminated fish (23). Only local families were supplied with seafood from the bay, because its small area, about 2 square kilometers, did not permit commercial fishing.

After fishing was banned in 1957, only fetal patients were subsequently reported for the next few months. This supported the theory that the poison was in the fish and shellfish. Dr. Hosokawa, director of the Shin Nihon Chisso Company hospital, had secretly confirmed this. After he reproduced the poisoning symptoms in cats, company officials kept the experiments secret and ordered them discontinued. In 1958 three new patients contracted the disease, and Mrs. Watanabe gave birth to a third and more severely afflicted child. With 16 more cases the following year, some doubts were raised that fish caused the disease since eating them had been banned. But the Kumamoto research team later concluded that the fishermen had continued to catch and eat fish and shellfish secretly, because even their very low standard of living could not be maintained without this major diet staple. Moreover, most victims became ill during the summer months when fishing was most extensive. However, other potential sources of contamination to Minamata Bay were also surveyed.

The waters were polluted by a slaughterhouse located in Tsukinoura, gushing water from the bottom of the bay near Yudo, ammunition discarded at the end of World War II, agricultural chemicals, and the Shin Nihon Chisso Company. This large chemical-fertilizer factory occupies most of the shoreline of Minamata city. South of the factory, mountains rise to form the bay and winding coastline. The Chisso Company was the major source of pollution and discharged its waste effluent directly into Minamata Bay (21).

Knowing their waste effluent induced the symptoms in cats, Chisso officials refused the researchers access to company property to conduct tests and thereby delayed confirmation of the cause of the disease. The drainage was finally checked when the factory rechanneled its waste discharge into the Minamata River in 1958. Until then the Kumamoto research team explored alternative possibilities, but when samples of the effluent became available, they quickly confirmed that a number of heavy metal compounds in the wastewater could be causal agents. The Chisso factory had poured its wastes into the Shiranui Sea before 1950 when the effluent was routed through a new drainage canal into the smaller Minamata Bay (see Figure 4). Tidal water does not readily flow in and out, and so it is a natural settling basin. Seawater remains in the bay for a long time, and the pollutants diffuse slowly into the Shiranui Sea. Consequently, after 1950 the wastes concentrated in the mud, seawater, and shellfish for 3 years before the first people were poisoned. Figure 5 shows the distribution of patients at Minamata (24).

After the drainage site was changed in 1958, contaminants were dumped for from 9 to 12 months before new patients contracted the disease near the mouth of the Minamata River. One person was stricken 9 months after moving into the area. However, just as the cause of illness seemed to be narrowed to pollutants from the Shin Nihon Chisso Chemical Company, people also began to develop symptoms as far away as Tsunagi to the northeast of Minamata City and across the Shiranui Sea at Izumi in Kagoshima Prefecture to the southwest. Authorities had warned the people at Minamata, but not the people at Izumi, not to eat the fish (24), and children died of congenital Minamata disease there as late as 1964. Later, the Chisso Corporation's claim was also challenged that no more of the waste

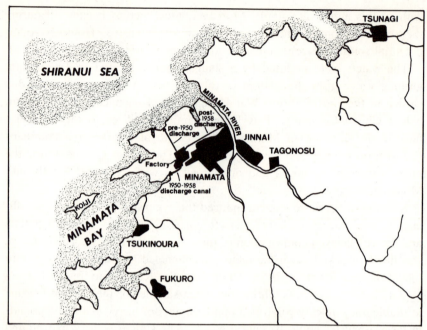

Figure 4. The wastewater discharge canals used by the Shin Nihon Chisso Company prior to 1950 (into the Shiranui Sea), between 1950 and 1968 (into Minamata Bay), and after 1968 (into the Minamata River).

had drained into the waterways after 1960. Other possibilities were also raised, such as that free-swimming fish like sardines could migrate from Minamata Bay to other localities and cause new outbreaks of the disease (21); or there might also be other sources of contamination (25).

Since the symptoms suggested heavy metal poisoning, a series of chemical analyses were conducted between October and December, 1958; and they confirmed that the factory discharged manganese, copper, iron, mercury, and lead as well as other elements and several organic substances. The heavy metals quickly settled in the bay near the plant's outfall, so levels were much higher in the mud near the pier and along the coastline than further offshore. The list of suspect elements was gradually narrowed by animal experiments in which the symptoms were compared with those from Minamata disease.

Manganese was quickly tested because an earlier poisoning epidemic had been caused by drinking well water contaminated with this element,

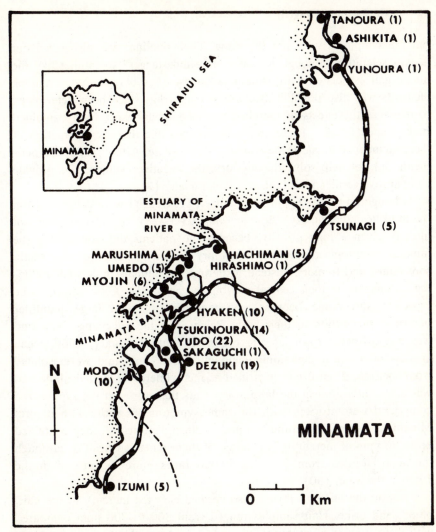

Figure 5. Distribution of Minamata disease patients according to village. Number of deaths are indicated within parentheses. From *Minamata Disease,* Kumamoto University, 1968.

but the symptoms were not the same. Then thallium was also ruled out as was selenium, although it caused a "blind stagger" something like the ataxic gait associated with Minamata disease. A synergistic effect among elements was also possible, because selenium has since been determined to decrease the effects of mercurialism in animals whose food contained both elements (26). Therefore, it is also possible that the interaction of these elements influenced the course of the disease. Although these experiments did not help solve the mystery, the negative evidence did provide comparative verification of the actual cause later (21).

Although up to 2010 ppm of mercury (wet weight) were collected from the mud near the factory's drainage outlet (27–28), this element was not initially given a high priority, because the patients did not display the familiar symptoms of inorganic mercury poisoning such as loose teeth, sore gums, and tremors. Instead, experimental animals underwent pathological changes in their central nervous systems (21). Cats seemed to be especially susceptible, because they consumed relatively large quantities of fish in proportion to their body weight. But the poisoning compound was not extracted from aquatic organisms until February, 1969, when crystals of a sulfur containing methylmercuric compound were isolated from shellfish. Then it was identified as methylmercuric methylsulfide and was also synthesized in the laboratory. Cats that were fed the synthesized compound were stricken with the same symptoms as those with natural Minamata disease. When the fish and shellfish were tested, they contained up to 50 ppm of mercury (29). Some of them remained healthy, although they concentrated from 5000 to 50,000 times more mercury than the 1 ppb in the water (30).

Tests on the human subjects also verified elevated mercury in the hair, blood, and urine. Hair samples ranged from 300 to 700 ppm compared with 1 to 3 ppm of mercury in normal subjects. Although mercury in hair was later recognized as a fairly reliable indication of poisoning, the levels had declined after people stopped eating contaminated fish at Minamata. When men with short hair were checked, they had as little as 4.3 ppm of mercury, whereas women's long hair contained more mercury in the sections farthest from the scalp. Autopsies also revealed excessive mercury in the brain, liver, and kidney (19). The diagnosis of methylmercury poisoning was confirmed when pathological changes in the brain structure were compared with those in an individual who died in

1954, 14 years after being poisoned while manufacturing alkylmercury fungicides (31–33).

Although the symptoms of alkylmercury poisoning were verified at Minamata, the origin was still not certain, since the chemical plant used only inorganic mercury compounds, and they usually concentrated in the sediments. To explain the conversion from inorganic to organic compounds, one theory proposed that organisms in the waterway could methylate mercury biologically. This idea was rejected when minute quantites of methylmercury were discovered in the factory's waste effluent. For each ton of vinyl chloride or acetaldehyde produced by the Shin Nihon Chisso Company, between 500 and 1000 g of mercury catalyst were expended in the wastewater (21, 27, 34). When the plant discharged an estimated 220 tons of elemental mercury between 1949 and 1953, undetermined but significant quantities of methylmercury were produced as an accidental side reaction of the industrial process. When the crude vinyl chloride was washed to rid it of impurities, the methylmercury in the washwater was pumped into the bay (16). More methylmercury was discharged in the sludge when the reactors were cleaned every 90 days.

Since both acetaldehyde and vinyl chloride are intermediates in the manufacture of plastics, their production had expanded steadily to meet the growing demand. The Chisso Company had begun to make acetic acid with an inorganic mercury catalyst in 1932 and later used these catalysts to manufacture octanol and dioctylphthalate. Production of polyvinyl chloride resin began in 1949, and the monthly output increased from 60 to over 1500 tons by 1959. Acetaldehyde production also expanded from approximately 400 tons in 1949 to 2500 tons in 1959.

By 1960 the Chisso Company produced more vinyl chloride than any other factory in Japan and several times more acetaldehyde than its competitors. The number of victims increased along with the output (see Figure 6). A correlation was also drawn between expanded production at the Kanose factory at Niigata and additional cases of nervous system poisoning. However, people continued to be poisoned after the plant closed, because chemicals remained in the sediments of the Agano River and were gradually released. At Ariake-cho the contamination sources were alleged to be either an acetaldehyde plant operated by the Nihon Gosei-Kagaku Company from 1946 to 1965, 8 years before the outbreak, or a chloralkali plant operated by the Mitsui Toatsu Company in Ōmuta

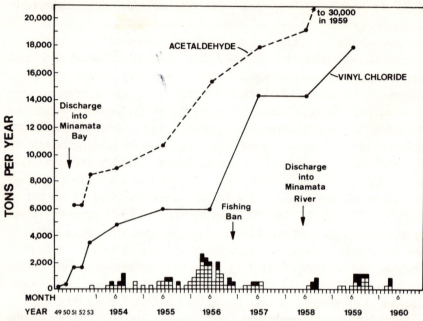

Figure 6. Correlations of the onset of Minamata disease and the acetaldehyde and vinyl chloride production at the Shin Nihon Chisso Company. Open and closed squares along the abscissa indicate the numbers of afflicted victims or deaths. Adapted from *Minamata Disease,* Kumamoto University, 1968.

(18–19). Generally, more wastes escape when a process is started, expanded, or halted, as these correlations indicate (22). However, even at peak times the quantity of methylmercury was small, much diluted, and only gradually assimilated by the fish.

THE HUMAN TOLL

As the cause of Minamata disease was traced, the symptoms were also more accurately described, but no cure has yet been found. Unlike acute inorganic mercury poisoning, much of the damage is permanent. Initial symptoms such as numbness of the lips and limbs resemble inorganic mercury poisoning, but the vision constricts with alkylmercury poisoning until the person seems to be looking at the world through a tunnel. Some 30 chemicals can cause visual constriction, so this in itself is not adequate to diagnose methylmercury poisoning. Observers first notice an irregular gait that resembles a drunken walk and signals a lack of muscular coordination. This may be the first sign that the central nervous system has been impaired. The loss of coordination makes it more difficult to swallow and chew as well as increasing the effort required to complete such simple tasks as drinking water from a cup, smoking, striking a match, buttoning, manipulating chopsticks, and writing. Continued physical deterioration causes a loss of speech, hearing, and taste. The speech may become slow, drawn out, and slurred. Unlike stroke victims who also slur their speech, Minamata patients have difficulty controlling their voices, so their speech becomes jerky and their faces contort in uncontrollable grimaces that resemble hysterical emotional reactions.

The heightening emotional disturbance could first be signaled by cycles of excitement and depression as well as insomnia (21). As the memory decayed, the victims often forgot how to write and spell (35). Then their emotions erupted into unpredictable, violent outbursts of anger or spells of depression. As their minds deteriorated, the patients went into convulsions and would laugh and scream without reason. As they became blind and deaf, sometimes their joints bent and stiffened into irregular positions, so the Japanese called them "living wooden dolls." Many of them had elevated blood sugar that indicated diabetes mellitus and changes in Langerhan's islet cells in the pancreas (20). At the terminal stage they became comatose and then died. Among the survivors the damage was originally thought to be irreversible, but over the years some people regained part of their lost eyesight, hearing, and motor coordination, because support cells replaced those that had been destroyed. However, after the gross symptoms were first delineated, other more subtle but no less frightening damage was also observed.

From the first, children seemed particularly liable to suffer serious residual effects. Their symptoms were often diagnosed as cerebral palsy, but too many babies developed this birth defect to be coincidental in families where other members had contracted Minamata disease. Up to 12% of the neonates had congenital defects in the small fishing villages compared with a national average of 5.8%. Both sexes were stricken, and by 1959 observation of 22 infantile cerebral palsy victims forced the conclusion that Minamata disease could also affect the unborn. Dr. Tadao Takeuchi, a Kumamoto University pathologist, confirmed this in 1961 and 1962 when he autopsied two children who had died of what he called "fetal Minamata disease." Whereas adult brain cells atrophied, in the fetus their development was arrested. The central nervous system could also be underdeveloped and malformed (36–37). Although smaller embryos might be so severely deformed that they would abort naturally, it appears that alkylmercury usually does not harm the fetus before the placenta forms (38). The brain's cellular architecture had formed normally in the children whom Dr. Takeuchi autopsied, and the bodies and extremities formed normally also. Therefore, the children were probably stricken in the intermediate or later stages of fetal development when the methyl-mercury disrupted the intracellular nerve cell differentiation at an advanced stage. However, in 11 of the 26 known fetal victims, the skulls were deformed as were the jaws and teeth (36–37).

With at least four times greater sensitivity to methylmercury, the fetus absorbed the compound from the mother and thus protected her from poisoning. Some newborn infants had 20 to 30% more mercury in their blood corpuscles than in those of the mother (20, 24, 39–40). Levels in the mother's blood and urine increased immediately after delivery. Mercury could also be passed to the infants through their mothers' milk, but no babies contracted Minamata disease while at their mothers' breasts, and later poisonings were often less severe than the congenital cases (20). Because the fetuses absorbed the mercury, some women had no symptoms other than elevated hair levels. They gave birth to apparently healthy infants who subsequently lacked enough motor coordination to sit up or crawl. When they were a few months old, some of these infants tended to throw their heads back in awkward positions and keep their legs crossed. The children who were this seriously damaged often died within 2 months of developing such symptoms. When the illness was more prolonged, they were susceptible to other diseases such as pneumonia, which sometimes became the primary cause of death. The survivors were often retarded and epileptic.

Since the mothers had no symptoms of poisoning, damage to the fetuses was dated indirectly from when the mothers ate contaminated seafood or from a month before other members of the family became ill, a time lag allowed for the latency period. One grandson's affliction was dated from when his grandmother contracted Minamata disease. And more fetal cases kept appearing. A 4-year-old girl with motor impairment and mental retardation was confirmed as a victim of fetal Minamata disease in February 1970, and the list totalled 26 by 1971 (36–37).

Besides the actual damage to the fetus when methylmercury crossed the blood-brain barrier, the specter was also raised that irreversible genetic damage could produce monsters or more subtle deviations in subsequent generations. Consequently, some parents refused to let their sons marry girls from Minamata. Although some severely afflicted girls who since have come of age had their menstrual cycle delayed, they do not seem to have suffered chromosome damage, and some have delivered normal babies as did other women who had given birth to fetal victims earlier. However, the life expectancy appears to be somewhat lower among women of child-bearing age around Minamata than in other parts of the country (20). And two children were born with rare deformities that usually occur at a rate of less than 1 per 1000 births. Although they cannot be traced to Minamata disease, the high percentage is cause for some alarm, and Swedish scientists have established that individuals who ate fish with elevated mercury levels had more of an uncharacteristic chromosome segregation than a control group (31). Moreover, methylmercury induced chromosome damage in the cells of onions as well as mutagenic effects in fruit flies (41–44).

Although the evidence is still incomplete on chromosome damage, other long-range implications are raised by what has been called "masked Minamata disease." After 1960 few people suffered acute poisoning, but each year more chronic poisoning is confirmed. In some instances the victims had no clinical symptoms and died of other causes. Then an autopsy confirmed that the brain structure had altered. For example, an autopsy confirmed that an old doctor had suffered tremors and mental disorder caused by Minamata disease. Sometimes the symptoms of central and peripheral nervous system disturbance persisted unchanged as mental imbalance and character abnormalities worsened.

Pathological degenerative changes, particularly in the white matter of the brain, sometimes were not clearly identifiable during the first 2 or 3 years but became more obvious after 4 to 6 years and even more widespread after 10 years. The weight of the brain was generally reduced, and gross changes included sclerosis, thinning of the myelin sheaths, and degenerative thinning of the cerebral cortex (20). Thus the aging process could be accelerated, because reserve cells were no longer available to replace the approximately 100,000 cells that the brain loses each day through normal attrition (45). At Minamata middle-aged people are showing signs of aging that are difficult to distinguish from the natural aging among their elders (38). Moreover, as other organs age, the body's declining efficiency also makes it more vulnerable to infections and such degenerative diseases as cancer, arteriosclerosis, and diabetes that often cause the deaths attributed to old age. Thus this subtle form of chemical poisoning may contribute to a general decline in life expectancy.

Slight contamination of the fetus may also induce subtle behavioral differences and decreased intellectual capacity. At the Minamata middle school pupils often seem to have a slight loss of coordination and peripheral senses, and a higher rate of feeblemindedness has generally been observed in that district (20). In fact, 38% of all children born between 1953 and 1960 appear to be mentally deficient, including the 8% known congenital cases. Subnormal mental capacity may be the only symptom of Minamata disease in some instances (46).

As more data are compiled on the symptoms and effects of methylmercury poisoning, new techniques are also being devised to diagnose subacute symptoms, to expedite excretion, and to rehabilitate the victims. Although a constricted visual field is characteristic of alkylmercury poisoning, approximately 30 chemicals induce similar reactions. But the disintegration of nerve cells in the cerebral postcentral-cortex causes several other neurophthalmological symptoms as well. The eyeballs may move abnormally; the short-sighted reflex may be disordered; and scattered, disintegrated dark points can be located in the center of the constricted visual fields. Children born with Minamata disease also have a lateral squint to the eye. To detect the ophthalmological symptoms such as the centripetal constriction of the visual field and the eyeball movements, the Kumamoto research team devised a neuro-opthalmometer (38). This instrument should be especially useful for mass diagnoses at clinics.

As research techniques were perfected to confirm the symptoms of methylmercury poisoning at Minamata, a pattern of secrecy and resistance was developed at all levels of society, from the victims through industry and government. As the victims' tragedy raised the possibility of less industrial growth if steps had to be taken to remove the contamination, the new antipollution laws were ineffective and allowed the source of poisoning to continue, recognized but unchecked. As Japan discovered and then struggled with this contamination from seafood, methylmercury poisoning was also demonstrated in other countries and circumstances.

2

FUNGICIDES AND PRESERVATIVES: ALKYLMERCURY SEED TREATMENTS

THE BIRDS WERE DYING

A decade after the Japanese disaster began at Minamata, Swedish farmers and conservationists became alarmed because increasing numbers of wild birds were seen flopping helplessly on the ground before they stiffened into irregular positions and died. This became so common that many species of birds were gradually disappearing from their old nesting habitats, and some were no longer found in former breeding areas. Among those that remained, sometimes both seedeaters and birds of prey refused to nest or even abandoned their nests before the eggs were hatched. Rotten eggs were often found in the nests of the white-tailed eagle and the eagle owl. The numbers of house sparrows and partridges declined as did the peregrine, kestrel, and hen harrier. These last three species live on mice and other small rodents that eat seed grain in the fields (47). Not only were birds in rural areas more often stricken than their city counterparts, but the illness seemed to be linked with the planting season for grain crops, because the first hatch of baby birds often died in the spring, whereas the second and third hatches survived (48–51). The birds' illness was traced to eating seed grain treated with alkylmercury fungicides such as Panogen[R] and Betoxin F[R]. A steadily increasing percentage of seed grain had been treated with these fungicidal compounds since World War II. According to later estimates, seed-eating birds and rodents managed to consume at least 1 of the 80 metric tons of alkylmercurial fungicides sown on Swedish crops between 1940 and 1966.

How much mercury the birds concentrated depended on the accessibility of the grain. The farmers' planting methods generally allowed approximately 1% of all the seed grain to remain uncovered on the ground in the spring; an even greater percentage was apt to be available to the birds in the fall, particularly if the season were so wet that the lumps of dirt did not cover the grain evenly. Exposure also depended on the width of the headlands left when the farmers lifted the hoes on the grain drills to turn around at each end of the field and the speed at which the grain was sown. It was also spilled from trucks during transport to the farms, and surplus kernels were sometimes dumped where wild birds could feed on them. The seed grain was usually exposed from 1 to 11 days before it germinated in the spring, whereas growth usually began within 5 to 29 days in the fall when a smaller crop was planted (52). Game birds that had ingested sublethal amounts of mercury-treated grain often had at least partially detoxified when the hunting season began later in the year. However, since alkylmercurials magnified as they passed up the food chain and people had been poisoned by eating fish contaminated with methylmercury at Minamata, the fear was raised that human beings could concentrate the poison not only by eating wild game birds such as pheasants and ducks but also from ordinary foods like eggs, chickens, and meat.

In 1965 Swedish eggs contained 4 times as much mercury as those from other European countries except Norway, where a methylmercury dicyanodiamide fungicide was commonly applied on seed grain (53). Eggs collected from small Swedish farms normally averaged 29 ppb of mercury. On a few farms where the levels went up to 1.5 ppm of mercury, the hens probably had been fed some seed grain that had been treated. At least one farmer admitted the practice, and others claimed they knew from long experience that the chickens' diets could include from 10 to 20% treated grain without noticeably affecting their general health or egg-laying capacity (54–55). With a diet of ⅛-part treated grain, the hens laid eggs with up to 4 ppm of methylmercury. Even if they ate no treated grain, however, the chickens concentrated some mercury when they ate grain grown from treated seeds. Their eggs averaged 22 to 29 ppb, whereas with no mercury exposure the levels were ⅓ to ½ as great (55).

As methylmercury passed from the treated seed grain to the crop and then through the hens to their eggs and chicks, presumably human beings at the top of the food chain were concentrating the element from several sources. With the breakfast eggs they might eat ham from hogs that had shared the farmers' leftover seed grain or corn with the chickens as well as bread made from flour containing fungicide residues. Some samples registered up to 3 ppm of mercury, and Swedish foodstuffs were generally higher than those from other European countries. Swedish ham cutlets, for example, contained 30 ppb of mercury compared with 3 ppb in Danish meat (53, 56–57).

Government action was necessary to control alkylmercurial fungicides in Sweden after Austrian and Danish markets refused to sell Swedish eggs because of their high mercury levels. In 1964 the Swedish Plant Protection Institute restricted alkylmercurial fungicides to the fall planting and grain used for scientific purposes with none allowed on grain planted in the spring. All alkylmercurial seed treatments were banned in 1966. Then preventive treatment of healthy grain was no longer allowed, and infected grain could only be treated with the less toxic alkyloxyalkylmercurial compounds, generally Panogen Metox[R]. With less mercury on the seeds and crops, the levels were also reduced in poultry, meats, and liver. Swedish eggs averaged 9 ppb of mercury in 1967 compared with 29 ppb 2 years earlier (49, 58–59). At the same time the wild bird populations began to return to normal. Whereas the number of rooks had declined drastically during the 1950s and early 1960s, they doubled over the 5 years before 1970 until their numbers again approached those of the mid1950s. Although bird populations had previously been declining faster in the country than in the city, the trend reversed after 1966. Wood pigeons in Malmo, Sweden then had 10 times more mercury in their livers than those in rural areas, because industrial pollution placed a greater mercury burden on the urban birds (48).

Although some farmers contended that a little treated grain was nutritionally beneficial (60), this folk wisdom ran contrary to the law in several countries. Although grain dressed with an organomercurial fungicide had been fed to hens and livestock without ill effects as early as 1922, England banned feeding cattle ingredients that were deleterious to their health in

1926 (61). Canada enacted similar legislation and imposed penalties ranging up to $1000 for mixing treated grain with market grain (62). But enforcement was impossible. Alberta provincial officials acknowledged that their half-million farmers could not be closely supervised. Poultry and farm animals could be fed treated grain intentionally or accidentally if contaminated seed bins, truck boxes, and grain augers were not carefully cleaned or if the waste remained accessible. And the side effects could be much greater than the farmers realized.

Although methylmercury fungicides prevented many diseases on plants and seed grain, their manufacture and application exposed workers and farmers to new and more drastic illnesses than those caused by inorganic or arylmercurials. Hypersensitive workmen frequently suffered red and blistered hands, but inhaling the fungicidal dusts caused more serious symptoms, permanent injury, and sometimes death.

In 1940 four out of sixteen workmen were poisoned despite wearing dust masks, goggles, and gloves as they worked in a well-ventilated fungicide-manufacturing plant (32). They were exposed to alkylmercurial compounds, mostly methylmercuric iodide, nitrate, and phosphate, for up to 4 months. After working with alkylmercurials for only 2 months, one of the men developed blisters from burns on the right forearm. He was transferred to work with inorganic mercury salts before more severe symptoms set in. All four men first underwent emotional changes. Their dispositions deteriorated, and then they gradually developed staggering gaits, clumsy movements, and tunnel vision. All their exposures culminated in some permanent damage. When Dr. Donald Hunter observed the four afflicted men in 1940, he issued a warning and advocated that such dangerous compounds be outlawed. However, his objections were overruled on the grounds that the poisonings were caused by careless handling, a lack of safety precautions, or inadequate ventilation.

Of the four the 16-year-old laboratory assistant made the most complete recovery, and two 33-year-old laborers regained the ability to perform unskilled tasks within 2 years. However, a 23-year-old workman who had been exposed to alkylmercurials for 4 months subsequently spent 1 year in the hospital and remained totally disabled. When he died of pneumonia 14 years later, an autopsy confirmed that some brain cells had been destroyed and the central nervous system was damaged (33). This 1954 autopsy enabled Japanese investigators to confirm that patients who died of Minamata disease had, in fact, suffered from alkylmercurial poisoning.

Because relatively short exposures to alkylmercury vapors can cause permanent damage, subsequent poisonings from their misapplication ranged from bizarre to simply tragic. In 1942 two young stenographers were fatally poisoned by alkylmercury vapors while they worked 4 and 6 months, respectively, in the office of a warehouse in Calgary, Canada, near a stockpile of 20,000 lb of diethylmercury fungicide. This compound is so volatile that it is not safely restrained by gas masks and so poisonous that inhaling small quantities may cause an infirmity that lasts for weeks or even months. After the tragedy the fungicide was moved to an even more hazardous site where a steam pipe directly overhead caused more vapors to concentrate because of the increased temperature. It was finally moved to a warehouse where explosives were also stored and no people worked continuously (63).

Treating and testing the grain also generated occupational hazards. Workers who treated seed corn left their respirators hanging on a post so that the alkylmercury from dust explosions and leaks in the exhaust ducts collected in them. Then the respirators would have been a greater hazard than breathing the elevated vapors in the room if someone attempted to wear them. Overall, few seed treaters were afflicted with alkylmercury poisoning despite the high levels during treatment, probably because natural ventilation was good in the autumn when the grain was usually treated, and the workmen were exposed for only 2 or 3 months. Their systems could detoxify between seasons. However, ventilation was not always as good at seed control stations where workers weighed and tested the grain. They sometimes reported such general symptoms of mercury poisoning as headaches and fatigue as well as an unusual taste that caused them not to enjoy smoking. One woman contended that she could tell whether the seed she weighed was treated by how fatigued she felt at the end of the day (64).

During the planting season not only the birds but sometimes the farmers suffered toxic reactions when they planted treated seed. In Sweden such incidents were documented in 1929, 1932, 1949, and 1954 (65). Some fatalities occurred among at least 20 farmers who contracted mercury poisoning over the 3 or 4 years before a list of such cases was compiled in 1949. One farmer became an invalid after sowing treated seed grain by hand on a warm day. First his fingertips became numb. Then he developed vertigo and a speech impediment. His condition continued to deteriorate until he entered the hospital 9 months after the exposure. Then the farmer's speech was unintelligible, and he was subject to laughing spells. Treatment with British anti-Lewisite (BAL) may have helped him regain the ability to walk with support, write his name legibly, and speak intelligibly. This problem is less likely to occur when farmers plant their crops with machines and the exposure is less intense. Nor are their families as apt to use the seed grain for cooking when the standard of living is high enough to provide alternative sources of food.

IGNORANCE, POVERTY, AND TREATED GRAINS

When treated grain was available for planting, not only wild birds but children were particularly apt to be attracted to the red dye that is supposed to warn of the hazard. Although they usually did not eat lethal quantities, up to a dozen people were treated for mercurial poisoning from this source each year in the Canadian province of Alberta. Pesticide poisonings approximated the number from tobacco and alcohol in 1971, but the effects have been more extreme when families included the treated grain in their food.

When four members of a Swedish family ate food containing flour ground from treated grain in the late 1940s, the 1-year-old boy developed a rash and a fever, became listless, and lost the inclination to sit up and crawl about. After he was admitted to the hospital, his father was also treated for mercury poisoning and referred to the psychiatric department of the medical hospital at Lund, Sweden. The mother showed no toxic symptoms, nor did the baby girl born a month after her brother entered the hospital. However, by 1951 it was apparent that both children were retarded. The girl was still unable to sit up, and the boy had only begun to crawl again. In 1952 two other children were poisoned by eating porridge made of treated grain, and the same pediatrics clinic reported other less severe cases of alkylmercurial poisoning (66). Although these were isolated occurrences in Sweden, in more impoverished, underdeveloped countries epidemics of mercury poisoning from eating treated seed grain were reported as late as 1972.

The human toll has been greatest in rural areas where poor farmers traditionally grind their own seed grain into flour. Often their other food supplies are inadequate, and they also consume proportionately more of the homemade bread or porridge. Epidemics from eating treated seed grain have been traced in farm communities 3 times in Iraq, Guatemala, Russia, and among more affluent city dwellers in West Pakistan. Rumors of other such epidemics have gone undocumented in Libya and Iran. In both Iraq and Guatemala previous epidemics were also diagnosed after alkylmercury poisoning was recognized.

The 1956 epidemic in Iraq was confirmed as alkylmercury poisoning only in retrospect after the 1960 disaster. However, knowledge of the two previous incidents still did not prevent a third and more widespread epidemic in 1972 after the hazards of alkylmercury poisoning had been given wide publicity. In 1955 and 1960 the Ministry of Agriculture supplied farmers with seed grain treated with E. I. DuPont's Granosan M[R], which contains an ethylmercuric compound. In 1955 the dressed seed was distributed in northern Iraq during the planting season from October through December. Between January and April 1956, more than 100 farmers and their families were admitted to the Mosul Hospital, and 14 deaths ensued (67). In the second epidemic dressed seed wheat was distributed in central Iraq in late 1959, and patients flowed into the Baghdad Hospital in March and April 1960. Although 221 persons were admitted to one hospital, the total number afflicted was not tabulated. Patients often spoke of deaths in their family or locality.

The diagnosis was complicated by the time lapse between the distribution of the grain and the onset of symptoms. Some farmers only began to make bread from treated seed grain after their own stocks were exhausted or they believed other persons had eaten it without ill effect. And the poisoning symptoms did not appear for a few weeks or months until the body burden of alkylmercury accumulated. Therefore, people were lured into a false sense of security and encouraged others to eat the grain (68). Since they had been warned against it, even patients at the hospital refused to discuss their illness, because they feared that they had committed a crime. However, since farmers in Iraq often eat their seed grain when their food supply runs short, the doctors checked for mercury in urine specimens and confronted the farmers with the evidence (69).

The farmers then admitted having eaten the dressed wheat for some time. They washed the grain first or mixed in larger quantities of untreated wheat or corn. Others first tested it on chickens whose symptoms were also delayed. If the flour were ground solely from whole-grain wheat coated with fungicide, each of the traditional, pancake-shaped loaves of bread could contain approximately 6 mg of fungicide, corresponding to 2.5 mg of metallic mercury (70). People became ill after eating three loaves of bread per day for 9 to 100 days. When the illness was identified, the Ministry of Agriculture broadcast warnings that brought several hundred more persons to hospitals in central and southern Iraq.

A year later, in February 1961, another alkylmercurial fungicide, Agrosan GN[R], a mixture of phenylmercuric acetate and ethylmercuric chloride, poisoned about 100 people in West Pakistan (70). Because wheat was scarce and the seed wheat was of higher quality, it was purchased for food in January and February 1961. Consequently, wealthy families were also involved. They suspected and more readily admitted the cause of illness, because only family members who ate bread made from the treated wheat were afflicted. Some families may have prevented disease by washing the wheat thoroughly. The symptoms of poisoning were not the same as those in Iraq, probably because the mercurial fungicide was a different compound. The causative agent was confirmed by tests in which birds and dogs experienced essentially the same reactions as the human beings (71).

The alkylmercurial fungicide Panogen[R] was the poisoning agent for members of 12 poor farm families in the Guatemalan highlands in 1965 (72). Of 74 family members, 45 became ill and 20 died, a mortality rate of 44%. As at Minamata, infectious encephalitis was first suspected and had been diagnosed in similar outbreaks at Panorama, Guatemala, in 1963 and 1964. But the symptoms were not quite the same.

Before alkylmercury poisoning was confirmed, insecticides, paints, and unsanitary living conditions were also ruled out as causes among the farmers who lived principally by cultivating corn and wheat for sale and home consumption. Most of the people were stricken during the planting season, and more males than females were poisoned, perhaps because the men ate more of the treated food. However, children were most seriously affected. Fifty percent of the victims were under age 10, and 75% were younger than 20. Again the victims were fearful that they had committed a crime and were ashamed to admit that their poverty was so desperate. But the poorest families were most severely afflicted, and neighbors and local authorities confirmed that they had eaten the seed grain. Nonetheless, some individuals shared the diet and remained healthy. Probably they either could tolerate more alkylmercury or were mistaken about how much they had consumed.

The largest epidemic from eating treated seed grain occurred in Iraq in 1971 and early 1972 after awareness of mercury's toxicity had been heightened in many countries. Although the American courts had suspended interstate shipment of treated seed grain, manufacturers were still permitted to dispose of warehouse stocks by export or by intrastate sales (73). In 1970 a complicated trade agreement was negotiated between the Iraqi government and Cargill Incorporated of Minneapolis, Minnesota, or one of its international subsidiaries, in which 73,201 metric tons of treated wheat and 22,262 metric tons of treated barley were shipped to Iraq. The barley came from the United States, but the origin of the wheat as well as where it was treated and shipped has been variously attributed to the United States, Canada, and Mexico. The grain was dyed red, and warnings such as the skull and crossbones as well as the word "Poison" were stamped on each bag in English and Spanish, presumably the languages of the country of origin. However, many illiterate Iraqi peasants would not have been able to read warnings printed in Arabic either, and they may not have been familiar with the symbolic red dye and skull and crossbones (74–75).

The potential for disaster was massive. Flour made from this treated grain averaged between 8 and 9 ppm of alkylmercury, so 1000 tons could poison 60,000 people. Since almost 100,000 metric tons of treated grain had been sent to an agricultural country of 10 million people, half the population could conceivably be wiped out as the grain was distributed to every province. The Iraqi government again broadcast warnings not to eat the grain, but many people in rural areas either did not have radios, did not believe the warnings, or ignored them because they lacked other food.

Timing was also a factor. Although the grain began to arrive in October 1971, some of it may not have been distributed until the end of the planting season, as late as January 1972. Where it was planted, birds died from eating exposed grain on the ground; but people again washed the dye off, tested it on animals, and then began to make bread. The Ministry of Health officially recognized 50 cases of poisoning in January 1972, and hospital admissions soon reached 400 per day, all people who made their own bread in rural areas. None were from cities where the flour for commercially prepared bread is government inspected.

Much of the barley was thought to have been fed to domestic animals. They soon began to sicken and were rushed to the slaughterhouses. Since local meat supplies were contaminated, sales had to be banned (76). With an epidemic underway the Iraqi government recalled the poisoned seed and decreed the death penalty for anyone caught selling it (77). Then frightened peasants dumped the seed in the Tigris River, and the sale of local fish was also forbidden because of potential poisoning. The government eventually confiscated 5000 tons of mercury-treated grain, but 75,000 tons remain unaccounted for (74).

The Iraqi government quickly instituted a news blackout when the epidemic began, and only brief official announcements were released in the state-controlled newspapers and radio broadcasts (77). Unofficial estimates indicated that up to 60,000 peasants could have eaten enough alkylmercury-treated grain to cause some damage (76), and many did not have access to the overcrowded medical facilities. Tourists reported that thousands of people suffered brain damage, blindness, and paralysis. The government officially acknowledged that 6530 victims were hospitalized and 459 died. They represented every province in Iraq. Most of the patients were admitted to hospitals in the provinces of Kirkuk and Ninevah to the north; Qadissiya and Muthanna to the south; and Babylon, Diala, and Waset near Baghdad. The frequencies of hospital admissions per thousand rural population are presented in Figure 7. They were of all ages, although 34% were under 10 years old.

Much information on this epidemic, particularly the long-range effects, will not be tabulated for years; but it is certain to provide a comprehensive study of direct alkylmercury poisoning. For one thing, despite the news blackout, investigators from the University of Baghdad who studied the previous epidemics quickly sent a brief, almost plaintive, note to the prestigious *British Medical Journal* saying an epidemic was in progress and requesting assistance (67). Although diplomatic relations between the United States and Iraq were strained, Dr. T. W. Clarkson of the University of Rochester was invited to bring a research team to Iraq because he had been experimenting with a new polystyrene sulfhydryl resin, 17-B, that promised to hasten the elimination of alkylmercurials from the human body. In laboratory experiments on mice the resin acted like a chemical magnet that kept the methylmercury in the excretory channels so it could not reabsorb through the blood and tissues to concentrate in the brain or

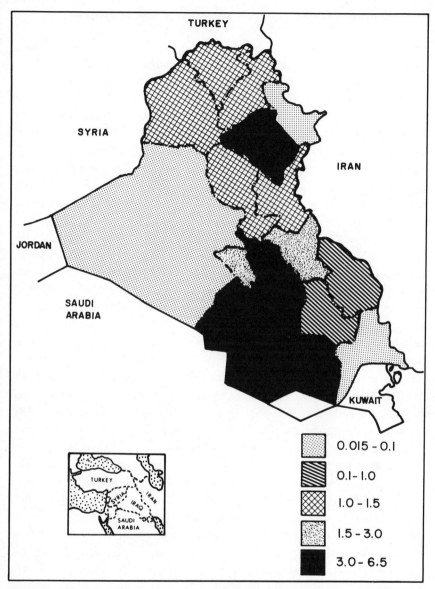

Figure 7. Hospital admissions per 1000 rural population in Iraq during the 1972 methylmercury poisoning epidemic. Courtesy of T. W. Clarkson, University of Rochester, from "Methylmercury Poisoning in Iraq," Bakir, F. et al., *Science* Vol. 181, pp. 230–241, July 20, 1973. Copyright 1973 by the American Association for the Advancement of Science.

39

central nervous system. Although it could not reverse damage already done, 17-B might decrease the severity of the poisoning (74).

Clarkson arrived in Iraq with enough resin to treat two people, neatly packaged in doses suitable for about 200 mice (78). The Dow Chemical Company agreed to rush more resin into production and to donate 100 lb to Iraq (74), and the Food and Drug Administration (FDA) acted in record time to authorize a permit to administer experimental doses (79). Despite the quick action, most of the resin arrived after the alkylmercury damage was done. Consequently, more was learned about the poisoning damage than the experimental cures, although some sulfhydryl resins and penicillamines were administered.

The sequence of poisoning stages was quite similar to those at Minamata. The lowest threshold body burden for the onset of numbness in the fingers and toes was approximately 25 mg of methylmercury. Ataxia began at 55 mg, difficulty speaking at 90, deafness at 170, and death at approximately 200 mg of methylmercury (75). Although 52% of the patients were female, only 31 pregnant women between the ages of 20 and 30 were hospitalized; but 70% of them died. Mercury levels were charted in these women and their infants, some of whom were born retarded. Since the new mothers insisted on breast-feeding their babies despite warnings, up to 2 ppm mercury could also concentrate in the infants' blood from the milk. In contrast, some foreign women in Baghdad were rumored to have had their babies aborted in fear that they might have been injured (75, 77–78). Although the damage is permanent, recovery by compensating cells was greater than expected. Some of the mildly or moderately afflicted patients gradually lost all symptoms. Even those who were more severely paralyzed responded favorably to physical therapy. They sometimes learned to walk again, and they regained other skills. Partial sight and hearing also returned in some cases without therapy (75).

While the epidemic was in progress in June 1972, Senator Gaylord Nelson addressed the Senate Commerce Committee on the Environment and raised the moral question of whether the American government should permit the export of products deemed too hazardous for domestic application. No action was taken, even though the domestic consumption of such fungicides had been banned. Sweden had also continued to export alkylmercurial fungicides and treated grain after their 1966 domestic ban (74). The Iraq disaster was also described to a world conference on pollution at Stockholm in the summer of 1972, but no agreement was reached on an international mercury export policy. However, unlike the tragedy of thousands of people in Iraq, the poisoning of one family in the United States had prompted domestic regulation of alkylmercurial seed dressings.

THE HUCKLEBY POISONING

In Alamogordo, New Mexico, 14-year-old Amos Huckleby, blind and in a wheelchair, was the first documented American victim of alkylmercury poisoning to receive widespread public attention. The ten-member Huckleby family included the parents in their 40s, three boys, three girls, and two small grandchildren. Ultimately, two of Amos' sisters and his baby brother Michael were also stricken. Ernestine Huckleby, age 8, was the first member of the family to become ill on December 4, 1969. She fell off the monkey bars at school and came home complaining of dizziness and pain. During the following week she began to stagger when she walked and grew steadily worse until she was hospitalized. Two local doctors tentatively diagnosed spinal meningitis or a blood clot on the brain, and she was referred to a neurosurgeon in El Paso, Texas. As her staggering persisted, her vision was gradually impaired, and she became more mentally agitated, no conclusive diagnosis was reached. Consequently, she was discharged to be observed as an outpatient. She was soon readmitted to the hospital and then went into a coma that lasted nearly a year.

Her condition was thought to be an isolated illness until two other members of the family were stricken. Amos, 13, developed a pain in his left side on the night of December 18, 1969 and then lost much of his motor control and part of his vision. Before he was admitted to the hospital in El Paso on January 11, 1970, he was uncommunicative, blind, and mentally agitated. Like his younger sister he subsequently became comatose. The day after Christmas, 18-year-old Dorothy Jean developed the same symptoms, and she was admitted to the hospital with Amos. In fear that an epidemic was developing, local and federal health officials were consulted, and the medical experts agreed that the disease symptoms resembled viral encephalitis, which develops gradually and infects the gray matter of the brain. But encephalitis is spread by ticks or mosquitoes, and there were none in New Mexico in December. Other possibilities were encephalopathy or heavy metal poisoning that might damage the brain or central nervous system.

An investigation of the family's living conditions revealed only one unusual feature. They had eaten a great deal of pork from a large boar slaughtered in September. Huckleby, a $281-a-month janitor at the junior high school at Alamogordo, raised the hogs to supplement the family's livelihood. Dr. Bruce D. Storrs later said the family might have eaten the meat as often as three times a day, although Mrs. Huckleby recalled that they ate beef occasionally as well (35, 80). Nonetheless, the pork looked and tasted like any other, and some members of the family had eaten it and remained healthy. Furthermore, no source of contamination was discovered at the butcher shop where the hog was slaughtered. However, 14 of Mr. Huckleby's feeder hogs had become ill a month after the boar was slaughtered. Huckleby later acknowledged that the boar had also shown signs of illness when he decided to have it killed.

Methylmercury poisoning was suspected after a ton and a half of grain, some a telltale pink, was found in a locked shed behind Huckleby's house. This was his share of approximately 5½ tons of grain sweepings, screenings, and chaff that he and five other hog growers had acquired from a grainery in Texico, New Mexico, in August 1969. The mixture of sorghum, oats, and other grains had been added to the hogs' ration of garbage and water. Then Huckleby cooked the mixture to comply with a New Mexico law requiring that garbage be cooked before it is used as hog feed. Huckleby kept the hogs in rented pens on a hog farm at the outskirts of town where he fed them varying proportions of the garbage collected from his neighbors. He added treated grain about once a week. The large boar was fed 5 gallons of the food per day, whereas the 14 feeder pigs were fed 3 gallons. Two sows and a gilt received 1 gallon each.

Mercury poisoning was confirmed by tests on the children's blood, hair, and urine. Hair samples ranged from 186 ppm for Mr. Huckleby to 2436 ppm for Dorothy Jean, the highest level ever recorded for a human being (81). Three other family members had elevated mercury levels, although they did not become ill. The two small children had little mercury, because they ate no pork. An older son escaped, because he was away from home at the time. Despite the high mercury levels, some confusion remained over the diagnosis, because the children did not display the classic symptoms of inorganic mercury poisoning. Methylmercury poisoning was finally identified by coincidence after Dr. Alexander Langmuir, Director of the Epidemiology Program at the Center for Disease Control in Atlanta,

Georgia, happened to read a 1960 account of Minamata disease and compared the Huckleby symptoms (82).

Meanwhile, doctors treated the three patients with BAL between January 19 and January 28, 1970. Amos and Ernestine did not respond to this or n-acetyl-d,l-penicillamine, an experimental chelating agent. However, Dorothy Jean improved and ultimately made the best recovery. Apparently, the mercury passed through her system before it could do too much damage. She learned to perform some household chores, read magazines, write letters, and follow television programs. Despite speech and motor difficulties, some hospital personnel at the rehabilitation center thought she was capable of doing college work and recommended that Dorothy Jean be enrolled at a 2-year school (81).

Ernestine and Amos were not as fortunate. They were placed in the chronic care facility of an Alamogordo hospital where their recovery was very limited. Ernestine regained consciousness, although she was still blind and could do little except roll over and move her arms a few inches to play with simple toys. Amos was also blind and partially paralyzed, although he could speak with some difficulty when he was discharged from the hospital in May 1971.

As the stricken children were treated, the 40-year-old mother was also carefully observed. Although she showed no symptoms of alkylmercury poisoning, she was pregnant when she ate the pork. Baby Michael was born on March 9, 1970. He appeared normal despite violent convulsions a few minutes after birth. The boy had intermittent tremors of the arms and legs for several days and a weak, high-pitched cry. Nonetheless, Michael's electroencephalogram test patterns were within the normal range at first. They later showed more erratic peaks that suggested brain damage. And the tremors became more pronounced even after he no longer excreted mercury (83). By his first birthday Michael was irritable to the point where he cried all day and refused to be separated from his mother. Medication calmed his nerves, but the child had an allergy to something in the desert air that made breathing difficult (81). He was also blind and retarded.

The family subsequently sued the New Mexico Mill Elevator Company and Ray Pritchett, the representative for the Golden West Seed Company, as well as the Nor-Am Agricultural Products Company, manufacturers of Panogen[R], and the parent company, Morton International, and its subsidiary, the Morton Salt Company (81). In 1975 the $3,925,000 suit was dismissed on the grounds that Ernest Huckleby had not heeded the warnings for the proper use of treated seed grain.

Some factors in the case still remained ambiguous. Both factory spokesmen and the hog growers were somewhat vague as to how the men came into possession of the treated grain. The men said they purchased some cattle feed and were given the sweepings. Factory representatives explained how the grain was initially screened to remove chaff and small kernels that were piled in a shed outside the grainery and disposed of periodically. This waste grain was what Huckleby and the others were supposed to have acquired. When the treated seed was dyed pink, stacked, and stored for sale, any that fell on the floor was swept up, bagged, labeled as poisonous, and stored in a locked warehouse to be burned. The sweepings accumulated during the summer and fall of 1969 because weather conditions did not permit burning. Company representatives acknowledged that the men might have acquired some of the poisonous sweepings. If they were piled on top of bags of untreated grain for the trip back to Alamogordo, the fungicide could have leached onto the feed grain when a heavy rainstorm soaked the bags. Or Huckleby and his friends might simply have disregarded the warning pink dye.

Neighbors ate with the Hucklebys occasionally but not often enough to accumulate much methylmercury, although one friend was given the hog's liver and kidneys, the organs that often concentrate the most mercury. But a larger public health issue was involved, because 132 potentially contaminated animals were sold to hog dealers between August 1969, and January 20, 1970. According to a Public Health Service report, the pigs were sold and processed through the Roswell, New Mexico, packing house as follows: 13 in August, 17 in September, 44 in October, 11 in November, 6 in December, and 44 in January. Probably the hogs sold in August would not have accumulated much methylmercury, but sales each subsequent month included hogs that had eaten more contaminated grain. However, more hogs were sold after the hog growers were requested to keep their livestock and to stop feeding them treated grain until the methylmercury could be excreted. The half-life of this compound had not been specifically determined for hogs, but elimination was expected to require at least 2 or 3 months.

The sales were discovered in January 1970, when Cade Lancaster saw the name of one of the hog growers listed in the recent sales records of a

Clovis, New Mexico, hog broker. Lancaster knew about the incident, because he had identified Panogen[R] as the chemical contaminant on the Huckleby grain, but the broker did not. He bought the contaminated meat at the Clovis purchase center 200 miles northeast of Alamogordo. The last group of hogs had been included in a consignment of 248 shipped to a packing plant at Roswell, New Mexico. They had been slaughtered before the plant was notified, so individual carcasses could not be identified. Consequently, all the pork was embargoed. Although subsequent tests revealed only one contaminated hog (80), it seems likely that some of the 24 escaped the embargo, as five out of seven of the remaining hogs had to be sacrificed because of excess mercury. They had been sold to a private individual. The other two were sold in Muleshoe, Texas.

Although assurances were publicized that no contaminated meat had been circulated, a spokesman for the U.S. Department of Agriculture would say only that the meat under detention might have come from the contaminated hogs, but "we wouldn't say at this time that we had all the meat for sure." And Bruce Storrs of the New Mexico State Department of Health and Social Service sent a letter describing the Alamogordo poisoning to all physicians and veterinarians in New Mexico. They were asked to contact the Preventive Medicine Section if any unusual central nervous system illnesses were reported (82).

Probably, the contaminated meat was distributed widely enough until individuals would receive only a limited dose, unless someone purchased a whole side of the pork. Only one mysterious individual mercury poisoning was publicized in the southwest in the summer of 1970, and it was not likely to have been related. In Jasper, Texas, a 57-year-old veterinarian had 36 ppm of mercury in his blood when he checked into the veterans' hospital. Over the previous 6 months he had become absentminded and restless. He gradually became sloppy and irritable as the pain increased. No source of contamination was discovered, but inorganic mercury poisoning was suspected because of the type of symptoms and the fact that a single dose of BAL reduced his blood mercury to 16 ppm, and subsequent injections lowered it even further (84).

The sale of Panogen[R] was halted the day after the NBC evening news televised Amos' return home from the hospital. Dr. Harry W. Hays, Director of the Pesticide Regulation Division of the Department of Agriculture, sent a telegram to Nor-Am Agricultural Products Incorporated in Chicago and ordered the company to recall its supplies (85–86). A clarifying letter sent the next day explained that the Alamogordo incident was only one of several that had been reported in California, Oregon, and Texas.

In some of the other incidents the people ate treated seed grain directly. In July 1969, Mexican authorities in the Baja California Public Health Service contacted the Imperial County Health Department at El Centro, California, to analyze wheat believed to have been treated with Cerosan[R] as well as some tissues from a dog. A month earlier a Mexican agricultural worker in Imperial County had been given a sack of seed wheat by his employer. He took the grain back to his home in the Mexicali Valley of Baja, where it was ground into flour and eaten by the family, a dog, and some chickens. After the chickens died and the children became blind and paralyzed, the dog was sacrificed to corroborate the diagnosis, since he had similar symptoms. The children also died before the end of July (87).

In October 1970, the California Department of Public Health was notified that a man in San Diego County was feeding treated grain to his chickens, and at least one family was using a cup of contaminated grain each day as part of a bean dish. Local health authorities warned the families of the hazard, and no illnesses were reported. The grain came from a 10-million-lb shipment that originated in Canada, was treated in Arizona, and then was shipped from San Diego to Syria. During transport to the coast in 75 freight cars, about 1050 burlap bags had ruptured. Although approximately 850 bags were salvaged, 26,000 lb of wheat were still missing. The FDA inspectors located about 50 lb in Tijuana, Mexico, where three families were adding it to their food, and another 150 lb were stored at the chicken ranch. Mexican authorities were contacted, and warnings were placed in newspapers and on television in Tijuana. Since no more grain was found, the FDA concluded that most of it had been sent to a local dump after cleanup operations at the Crescent Wharf in San Diego during the week of October 12–16, 1970 (88).

The shipment of treated grain to foreign countries generated very little interest in the American press, and only the Huckleby poisoning in Alamogordo was given national publicity. However, when contamination was discovered to be causing elevated mercury levels in fish, a national issue emerged in both the United States and Canada. By then, more was known about waste discharges of the element into waterways.

3

NATURAL METHYLATION:
A SWEDISH DISCOVERY

The mounting accumulation of mercury from industrial waste discharged into the waterways caused little concern until the 1960s. The element and its compounds were thought either to remain stable or to react with other elements in the waterway and form relatively harmless compounds. This theory was refuted after Swedish scientists realized that the fish primarily concentrate methylmercury downstream from pulp and paper mills that released phenylmercuric acetate. The possibility of a natural conversion from one form of mercury to another in the waterway had been raised at Minamata a decade earlier and then rejected when methylmercury was determined to be a byproduct of the industrial process (16, 53, 89–90).

Downstream from a pulp mill on a river blocked by a dam, pike concentrated 5 to 10 times more mercury in their tissues than the fish upstream. Between 1964 and 1966 the range was from 0.16 to 0.83 ppm above the dam and from 1.5 to 3.1 ppm downstream. After the mill halted its mercury discharges in June 1965, the pike downstream still escalated from 3.4 to 9.8 ppm, apparently because the fiber deposits remaining in the river continued to release mercury (47).

Enough mercury had collected in Swedish waterways by 1966 until over 0.20 ppm were common in marine fish from coastal waters. In contaminated inland lakes and rivers freshwater fish generally ranged from 0.4 to 1.0 ppm of mercury (Figure 8). Moreover, 1 to 5 ppm measurements

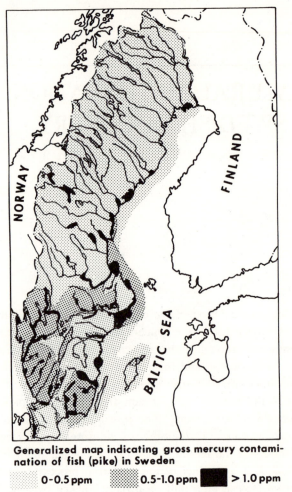

Generalized map indicating gross mercury contamination of fish (pike) in Sweden

0-0.5 ppm   0.5-1.0 ppm   > 1.0 ppm

Figure 8. The extent of mercury-contaminated pike in Swedish waterways in 1968. Courtesy of Göran Löfroth, *A Review of Health Hazards and Side Effects Associated with the Emissions of Mercury Compounds into Natural Systems.*

were frequent, and in 1969 the highest level, 9.8 ppm, was recorded for a pike taken downstream from a pulp mill (48, 91–93). Although the fish might also absorb or ingest other mercury compounds, the methylmercury was more readily assimilated and remained much longer in their bodies.

Methylmercury readily gravitates to the red blood cells and then to the fatty tissues of the fish. It is at least 1000 times more soluble in fats than in water and concentrates in the muscle tissues, brain, and central nervous system (94). Whereas phenylmercury has a halflife of less than 6 months, fish retain methylmercury for up to 2 years because it only gradually degrades into inorganic mercury to be excreted (95). The concentration rate depends on the duration of exposure as well as the age, size, species, and individual sensitivity of the concentrating organism. Apparently, if all other factors are equal, feeding habits and metabolic rates influence mercury accumulation more than the location of the fish on the food chain.

Bacteria were immediately suspected as the causative agents in the natural methylmercury conversion, because they are able to alter mercurial compounds in several ways. Some of them adsorb the mercury on the outside of their cell walls where it converts directly into a vapor. Others such as *Escherichia coli* absorb the mercury into their systems. Then the mercuric ion may combine with cytoplasm and degrade into a different compound or the volatile elemental form (96). The common *Pseudomonas aeruginosa proteus* and at least two other microorganisms were known to convert mercuric ion into elemental mercury by 1964 (97). In 1966 Gunnel Westöö confirmed that enzymes in liver homogenates could methylate inorganic mercury biologically (98). Then Jensen and Jernelöv demonstrated that organisms in bottom sediments could also methylate inorganic mercury (99–101).

More methylmercury is present in the water nearest the surface layers of sediments where microbes usually live. They apparently remove inorganic mercury that has adhered to food particles or dissolved in organic matter by methylating and discharging it into the sediments or water (102). Mercury buried in deeper layers of the sediment is not available for methylation unless it is disturbed. However, besides shifting currents and other physical phenomena, small worms can stir the mercury-rich layers to a depth of 2 cm, whereas larger freshwater organisms such as mussels can churn up the sediments to a depth of at least 9 cm (103). In places where ships drag the bottom of waterways, more than a foot of sediments may be stirred, and dredging to deepen the channels can disturb hundreds of tons (104).

Among sediment samples taken from more than 100 rivers and lakes in Sweden, all contained some microorganisms that were capable of converting other compounds into methylmercury (105–107). From the clearest natural waterways to anaerobic sewage treatment plants, microbial species are present to degrade other mercurials into inorganic mercury that can be methylated (108–109). The available oxygen supply determines which microorganisms methylate the inorganic mercury. In aquaria experiments oxygen-demanding microorganisms reacted the same as in the lakes and rivers where they normally live under similar conditions of light and temperature. When the surface layer of sediment was oxygenated in the aquarium, more mercury was available and the microbes increased their output of mono- and dimethylmercury, the two forms produced in nature (105).

Methylation may also occur in other ways. For instance, the fungus *Neurospora crassa* methylates mercury, and liver homogenates of tuna fish favor the biological methylation of mercuric ion. Some fish also appear to methylate mercury in the mucus on their scales and in their microbial fauna. Among warm-blooded creatures, liver homogenates of chickens and oxen evolve methylmercury from inorganic mercurial derivatives. The gut flora of birds and mammals are also likely to methylate mercury in their microbial systems, although the extent of this methylation has not been explored (110). Methylpentacyanocobaltate (a model compound present in Vitamin B_{12}) can also react with mercuric ion to produce methyl-mercury chemically. Since methylcobalamin is widely distributed not only in microorganisms but also in calves' livers, blood plasma (111), and the the kidneys of hogs (112), they also have the potential to methylate mercury chemically if not biologically.

Since widely varying conditions affect methylation even between one zone and another in a single watercourse, limiting or even predicting how much methylmercury will be released to concentrate in fish becomes a very complex process. Some mercury will still be methylated even if the organic pollutants and microbe population can be decreased. However, many waterways loaded with mercury have also been enriched by fertilizers and carbonaceous wastes such as wood pulp and sewage. The abundant food supplies from agriculture, industry, and municipal wastes have accelerated the growth of plants that feed the microbes. Thus although some mercury must have always been methylated naturally, the process has been accelerated by human intervention and is not apt to be reversed soon.

When organic pollutants enrich the waterways, the nutrients feed aquatic plants, and the microbes thrive and methylate more mercury. For instance, on a 26-mile stretch of the St. Clair River downstream from a chloralkali plant discharge site at Sarnia, Ontario, the highest rate of methylation occurred where a city without a municipal waste treatment system dumped its wastes into the river 8 miles downstream from the point of initial discharge (113).

Excessive organic enrichment may encourage the aerobic microorganisms to reproduce until their abundant population depletes the oxygen supply. Then they are replaced by anaerobic species that do not require oxygen. Moreover, both anaerobic and aerobic microbes may be present in the same watercourse under different conditions. For instance, some lakes stratify into two separate layers during the summer, and the surface layer (epilimnion) may be aerobic, whereas the lower layer (hypolimnion) is often anaerobic. Streams, rivers, and shallow lakes that do not stratify may gradually decay from aerobic to anaerobic as the oxygen supply is depleted. Then methylation continues under different conditions over the natural succession of phases during the life span of the watercourse (102, 114). A summary of methylation processes in the aquatic environment is presented in Figure 9.

Under controlled conditions in the laboratory, more mercury appears to be methylated in anaerobic than aerobic environments. But the reverse may be true in natural waterways. Under anaerobic conditions hydrogen sulfide may be available to react with some of the mercury and form the highly insoluble mercuric sulfide, a compound that is quite resistant to biological methylation. However, if the sediments are aerated, the mercuric sulfide may revert to a water-soluble mercuric sulfate that is more easily methylated (105). Thus another contributing factor besides the kind and quantity of microbes in the waterway is the amount and chemical form of the available mercury.

In anaerobic sewage lagoons where excess mercuric ions were quickly precipitated as either mercuric sulfide or other mercury salts that adsorbed onto particles, the aquatic environment was less toxic to microorganisms (106, 115). Consequently, more of them survived to methylate a higher percentage of the remaining inorganic mercury. As some bacteria have their growth inhibited by excess mercury, it is not yet certain whether a minimum threshold quantity has to be present before aquatic microorganisms begin methylation. However, one strain of fungi, *Neurospora crassa*, did not methylate inorganic divalent mercury when the concentration was less than 10 ppm in a methionine limited substrate or under 1 ppm in a

TRAGEDY
Human beings, fish eating birds and other mammals eat the fish, concentrate the methylmercury further and may be poisoned quickly or very gradually.

CONCENTRATION
Very small quantities of mono- and dimethylmercury released to the water by the microbes are absorbed and ingested by the fish to concentrate to steadily higher levels.

METHYLATION
Microbes on the surface of the sediments methylate elemental mercury in anaerobic conditions and mercuric ion in aerobic waters.

Living microbes methylate mercury biologically while components of dead microbes methylate it chemically.

AIR
Evaporation and precipitation cycle naturally to the atmosphere from land and water with industrial pollution added.

LAND
Weathering, land runoff and industrial waste are discharged to the waterway.

WATER
Mercury compounds degrade into one of three forms: Elemental, Mercuric ion or Mercurous ion which converts into one of the other two depending on the pH of the waterway.

SEDIMENT
The mercury adsorbs on particles and sinks to the sediments where it is stored or methylated.

NATURAL METHYLATION

Figure 9. Natural methylation of mercury in waterways.

substrate with a great surplus of methionine (103). In experimental conditions aquatic microbes produced substantially more monomethylmercury when the concentration of inorganic mercury was increased from 1 to 10 ppm. Above that quantity the rate of methylation decreased. Since the progression was not linear, however, it appeared that methylating and demethylating organisms were in competition.

As little is yet known about the processes of biological methylation, the reverse is an even greater mystery. However, some microorganisms like *Pseudomonas K-62* can decompose organomercurials into metallic mercury, which then vaporizes into the atmosphere (116). Similarly, rats can convert methylmercury into inorganic mercury in their brain cells, and in dead fish the monomethylmercury sometimes converts into dimethylmercury, which then may diffuse as a gas (105).

4

LAKE ST. CLAIR:
AN INTERNATIONAL INCIDENT

Norvald Fimreite, though not the first scientist to discern that mercury, fish, and man are linked in the food chain, will almost certainly occupy much the same spot in the environmental revolution as Paul Revere does in the American Revolution. Mr. Fimreite sounded the alarm, loudly, in March, 1970 (117).

Public attention was focused on mercury contamination of Canadian waterways in 1969 when the Canadian Federal Department of Fisheries and Forestry embargoed commercial fishing catches from Lake Winnipeg, Cedar Lake, the Saskatchewan River, and the Red River in the province of Manitoba (118–119). The Canadian government set a temporary 0.5-ppm action level and decreed that all fish with more mercury were unsafe for human consumption. Initially, more than a million pounds of fish with from 5 to 10 ppm of mercury were confiscated and destroyed. Among the species detained were walleye, northern pike, bass, and jackfish. Thereafter, all fish taken from the Saskatchewan River were embargoed and tested for mercury. The Canadian government authorized the Fresh Water Fish Marketing Corporation to continue the purchase of fish to encourage commercial fishing in marginally contaminated waters like the Lake of the Woods (120). Based on selected samples, the "clean" fish were marketed and the others destroyed.

While mercury contamination was being studied in Saskatchewan in February 1969, Norvald Fimreite also projected that mercury-contaminated fish would be found in the waterways of Ontario and Quebec provinces, where two-thirds of the Canadian chloralkali plants were located. Fimreite's conclusion was based on answers compiled from questionnaires on which company officials estimated that 0.5 lb of mercury was discharged into the waterways for each ton of chlorine produced or approximately 200,000 lb of mercury per year (121). Fimreite's findings were substantiated in Ontario province when the Water Resources Commission surveyed mercury losses from pulp and paper mills as well as chloralkali plants. Six of them reportedly resupplied their mercury cells with an average of 44,000 lb per year, and provincial authorities estimated that they released between 25,000 and 102,000 lb of mercury to the watercourses each year. In addition, 8 of the approximately 40 pulp and paper mills still used mercury-based slimicides. Out of 48,000 lb consumed per year, approximately 3000 were released to the waterways. Half of this was assumed to be recycled to make pulp and paper. Therefore, these preliminary findings indicated that Ontario's watercourses could be receiving from 26,000 to 104,000 lb of mercury per year from these two sources alone (121).

In the fall of 1969 both Fimreite and George Kerr, the Ontario Minister of Energy and Resources, sent samples of fish taken from Lake St. Clair to a California laboratory for analysis. Instead of releasing their results, provincial officials notified the Dow Chemical Company of Sarnia, Ontario, of their rising concern about trace metals in fish. Then more tests were conducted, and Dow was notified that fish were above the accepted tolerance level downstream from the Sarnia plant. The company's vice president, C. B. Branch, later informed the Hart Subcommittee investigating mercury contamination that the company started a crash cleanup program the next day. Branch claimed the company had begun to monitor this element in the plant's outfall on July 4, 1969 (122). Then the two mercury cell plants had averaged daily losses of about 70 lb in a range from 47 to 195 lb. Between 1950 and 1970 losses had averaged 30 lb a day with a maximum of 200 lb on occasion.

When the International Joint Commission's Advisory Board met on February 11, 1970 to discuss pollution control in the boundary waters, the Canadians warned the Americans about the high mercury levels in fish taken from Lake St. Clair. Then Michigan officials also began formulating plans to survey mercury losses from chloralkali plants, paper mills, and other sources in the state (123).

The fish-contamination show did not "go public" until Fimreite's findings were published in a London, Ontario, newspaper on March 19, 1970 (124). Then the Canadian Food and Drug Directorate formally advised the American FDA of the mercury problem (125), and within a week 18,000 lb of walleyes were seized from commercial fishermen on the Canadian side of Lake St. Clair. The next day the Canadians banned all commercial and sport fishing on that lake. The American Embassy at Ottawa finally informed the Secretary of State about the mercury problem by telegram on April 2, 1970. By then, federal, state, and provincial officials on both sides of the border were feverishly studying mercury contamination while the newspapers screamed for action (124).

When Ontario provincial authorities banned the sale of perch and pickerel from Lake St. Clair, Michigan public health officials proclaimed that the news of mercury in the Great Lakes had burst upon them like "a thunder clap" (126). But the rumblings should have alerted officials on both sides of the border much earlier.

The U.S. Public Health Service contributed support to the Kumamoto University research team at the beginning of their study of Minamata disease in 1956 (127), and two years later *The Lancet*, a prestigious British medical journal, reported that the toxic substance might be in fish (23). In 1960 *World Neurology* published Dr. Leonard T. Kurland's description of the symptoms of Minamata disease. He also compared the mercury levels in Japanese shellfish with those taken from Galveston Bay, Texas, and Chesapeake Bay, Maryland (16). The Japanese and American chemical companies utilized different industrial processes, and the Galveston firms dumped their mercury into settling ponds rather than directly into Galveston Bay. However, enough mercury leached into the Texas waterway until the oysters had concentrated up to 1 ppm. Mussels in the Chesapeake Bay were as high as 2 ppm, above natural background levels, although still insignificant compared with the 30 to 50 ppm in shellfish in Minamata Bay. But that small inlet was far away, and American public health officials later indicated that they had not seen any of the early publications about the epidemic.

Nor had any Japanese representatives been invited to the Stockholm Pesticides Conference in 1966, although environmental mercury contamination was a major topic of concern. American and Canadian delegates had expected more attention to be focused on DDT, dieldrin, aldrin, and other organochlorine insecticides that were the current scientific rage back home (128). However, Dr. Lucille F. Stickel later recalled that the American delegation had taken samples of ducks and osprey, but no fish, to be checked for mercury in the Swedish laboratories, because no American scientists were performing equivalent analyses at that time. The Americans were reassured by the low mercury levels in the specimens collected from Atlantic seaboard estuaries (129). However, in 1967 the Federal FDA conducted a pesticide diet study in which five food samples were measured for each of six regions in the country. Although some of the 30 samples contained up to 0.05 ppm of mercury, the FDA concluded that no in-depth analyses were required (125).

At the 1968 University of Rochester conference on the toxicity of persistent pesticides, Dr. Lucille Stickel referred back to the Swedish analyses 2 years earlier and acknowledged:

> Beyond all question, we should do a great deal more on the mercury problem in the United States, but at least these data do not suggest that we have a great bombshell ready to burst here (129).

As Swedish scientists reported on their research findings since 1966, the American participants still looked for reasons why their circumstances would be different. The Swedish mercury problem might be caused by the northern climate, they speculated, or by archipelagoes that kept the water from circulating. They also wondered if mercury had been used longer in the chloralkali industry in Sweden. Dr. Alf Johnels assured them that Sweden had adopted mercury applications from American industry: inorganic mercury in the 1920s, phenylmercury in the 1930s, and alkylmercury in the 1940s (130). He also cautioned that the earlier tests on birds could lead to false conclusions about actual mercury levels in the United States. Since fish downstream from pulp mills commonly accumulated mercury and nearly all Swedish lakes and rivers contained microbes that could methylate inorganic mercury, it seemed likely that similar conditions might exist in the United States (131–132). He could easily be proven right.

Figures were readily available on American mercury consumption because the Bureau of Mines had published production and consumption data by industry since 1850. The element was closely monitored as a valuable and relatively scarce commodity that could cause a hardship if a shortage ensued, particularly in wartime. In 1967 when Sweden lost 40,000 lb of mercury to the environment (133), American industry and agriculture consumed 5.6 million lb annually with an estimated 25% or 1.4 million lb lost to the environment, 35 times more than in Sweden. That country had eight chloralkali plants with mercury cells in 1967, whereas the United States had 26. In 1968 the American chloralkali industry consumed over 1.3 million lb of mercury per year and projected an increase of 275,000 lb annually by 1975. Out of 74 plants producing chlorine in 1969, 38 had mercury cells (134).

The implication was clear. The Americans were not finding mercury in the environment because they were not looking. With new, sophisticated analytical methods, the Swedish scientists had looked for and found methylmercury in fish (135–136), and it could not be removed from either the fish or the waterways. Contaminated fish retained the mercury long after they were placed in unpolluted waters. Thus Arne Jernelöv predicted that the mercury levels would be lower only in a new generation of fish after the mercury was no longer in the waterways (137).

When the mercury levels in fish taken from Lake St. Clair were publicized, one U.S. government official doggedly declared that "we see no imminent danger, but we're concerned enough to extend the investigation" (138). The FDA also belatedly advised their district offices and all state governments that mercury contamination could be a problem in the waterways and recommended methods to test for mercury in fish.

Both the United States and Canada previously had zero tolerance levels for mercury in food. Since the FDA had ruled that mercury was an unintentional food additive, if the law were strictly followed, all fish with any mercury residues, in other words all fish, could have been seized. Instead, the United States followed Canada's example and both countries quickly adopted an interim 0.5-ppm "action" guideline. Fish that exceeded it would be seized and destroyed. At the same time, no fishing would be permitted in lakes and streams where the fish consistently registered above the 0.5-ppm interim guideline.

The 0.5-ppm standard and a 10-fold rather than the usual 100-fold safety margin were justified on the grounds that the Japanese consumed up to 200 g of fish per day or 5 times the American daily average of 40 g. Moreover, fish from Lake St. Clair averaged around 2.5 ppm, 1/4 the Japanese average. The FDA safety margin was clearly based on averages rather than possibilities, since some fish in Lake St. Clair have tested up to 7 ppm, and people with access to fisheries often consume more than the national average. An estimated 0.1% of the American population, some 200,000 persons, eat more than average quantities of fish because of choice, opportunity, or necessity.

The Canadians also justified their 0.5-ppm interim guideline on the basis of average fish consumption instead of the dietary habits of minorities, such as the Walpole Indians who live on a reserve beside the St. Clair River or the Ojibway Indians who live on White Dog and Grassy Narrows reserves along contaminated rivers in Ontario province. But much publicity was given to individuals like Diego Barraso, a Canadian commercial fisherman and tugboat captain, who contended that eating mercury-contaminated fish made him stronger (139). Similarly, one family insisted that none of the six members suffered ill effects from eating fish taken from Lake St. Clair 3 times a day during the winter of 1969–1970 (140).

In contrast to these testimonials from personal experience, Canadian and American health authorities indicated that no one really knew how much mercury an individual could tolerate safely, but a total diet with 0.5 ppm mercury would be cause for concern. When a member of the Hart Subcommittee asked Dr. Albert C. Kolbye, Jr., of the FDA if he thought eating fish with 0.5 ppm of mercury was safe, Kolbye replied, "I would prefer to state that eating fish with 0.5 ppm levels of mercury certainly is far safer than eating fish with a higher level of mercury" (125).

To reassure the public that a little poison was all right, 0.5 ppm was equated with one shot glass full of vermouth in a railroad freight car full of gin. In a very large martini $\frac{1}{2}$ a bottle of vermouth would be in a 0.5-ppm ratio with 1 million bottles of gin. By the same token, only $\frac{1}{2}$ lb of mercury could contaminate 1 million lb of fish to the FDA action level (141). However, some fish in Lake St. Clair already registered up to 7 ppm. Moreover, as steadily more mercury had been dumped into the waterways over the years, the levels in fish appeared to be rising if samples from preserved specimens were accurate. They had been taken from Lake St. Clair, Lake Erie, and the lower Detroit River between 1927 and 1964. Measurements of mercury in the old fish could be inaccurate because of losses from vaporization, additions from preservatives, and a limited and perhaps unrepresentative number of samples, but the trend was still generally upward. For instance, in 1937 yellow perch ranged between 0.03 and 0.17 ppm of mercury, whereas the same species was 10 to 20 times higher in 1970 (142). Nonetheless, Dr. John Wood, who discovered some of the mechanisms of natural chemical and biological methylation, estimates that no more than 100 lb of inorganic mercury, less than 0.2% of the total deposited thus far, has been methylated (143). Thus the remaining tonnage can be gradually methylated and concentrated by fish and other aquatic organisms over the next century.

The rising mercury levels seemed of less consequence as the awesome economic implications became more apparent after Canada also banned commercial fishing on Lake Erie when walleye and yellow perch were determined to exceed the 0.5-ppm action level. As one state after another followed Canada's lead, fish sales plummeted throughout the midwest, and the Lake Erie and St. Clair commercial fisheries became an economic disaster area. On Lake St. Clair commercial fishing was banned entirely, and the fishermen had been loaned $100,545 by June 4, 1970, with no indication when they would be able to fish again and pay the money back (144). As the sport-fishing business also slumped when fishermen heeded public warnings not to eat the fish (145), Canadians and Americans were faced with enormous economic losses if the ban continued.

The cost to the sport-fishing industry could tally in the millions of dollars on each side of the border. In Michigan where thousands of sport fishermen seek recreation near the populous metropolitan Detroit area, the fishing ban put a damper on numerous tourist camps, boat rentals, and bait shops. The 97,000-acre Lake St. Clair had supported 200,000 angler days of fishing per year or about 2 days per acre in 1970 and was estimated to be capable of supporting 10 or more angler days per acre per year. Thus mercury contamination could destroy a sport-fisheries industry with a $3,336,000 annual gross income in 1970 and an estimated potential gross value of $15,680,000. Moreover, the facilities and equipment that supported sport and commercial fishing were then estimated to be worth from $67 to $415 million (146). The price per pound of fish killed by pollution had been calculated from the cost to rear them in commercial hatcheries as well as the species' size and role in the ecological balance. Thus the Pollution Committee of the Southern Division of the American Fisheries Society estimated that the fishing ban cost sportsmen prizes in a range from sturgeon at $50 per lb through muskellunge at $10, down to bait fish at 3¢ per lb (147).

Faced with these enormous financial losses from sport-fishing revenues, pressures soon mounted for a more liberal policy. Then the Ontario provincial official reviewed the total fishing ban and instituted the first significant compromise, a policy called "Fish for Fun." The Americans followed with an equivalent catch-and-release program. Then the tourist camps could remain open, because fish could still be caught but not eaten. Although this new option was publicized before the normally crowded Memorial Day weekend in 1970, boat rentals were down nearly 100% on Lake St. Clair, few fishing licenses were sold, and the fishing camps were nearly empty (148). Later in the season the provincial government encouraged the Fish for Fun policy by offering the tourist camp operators loans of approximately 35% of their gross revenues to stay open (144). A compensation plan was also considered but never initiated for bait dealers, gas stations, and other businesses indirectly affected by tourism. However, this proved to be unnecessary, because business gradually returned to normal after mercury pollution was no longer front-page news. In fact, despite the initial economic decline, the policy modifications enabled the American states to collect a record $192 million, $9.3 million more than in 1969, for the licenses, tags, permits, and stamps that finance state fish and wildlife conservation programs (147).

Meanwhile, more states found contaminated fish in their waters. In addition to banning sport and commercial fishing on Lake Erie, New York state soon closed Lake Onondaga to fishing and placed limits on fish consumption from Lakes Champlain, Erie, and Ontario as well as three rivers: Oswego, Niagara, and St. Lawrence. However, as sport fishermen heeded the warnings not to eat fish, the tourist business slumped (145). Then it became steadily less likely that the discovery of mercury contamination would be given wide publicity or be followed by stringent fishing bans to protect public health.

5

THE AMERICAN STORY

OCEAN FISH

Whereas mercury contamination was thought to be a local problem in Lake St. Clair when Philip Hart's Senate Subcommittee opened hearings there in May 1970, this certainly was not the case. Other states outside the Great Lakes region soon announced that fish in their waterways also exceeded the 0.5-ppm recommended guideline. In California, Idaho, Oregon, and Washington excessive mercury was found in 80% of the brown bullheads that were tested, 74% of the northern squawfish, 54% of the channel catfish, 47% of the largemouth bass, and 11% of the white sturgeon (149). Moreover, mercury was a factor in a fish kill in New Jersey, and in Texas oysters from Lavaca Bay were contaminated with up to 5 ppm of mercury (150). Before long the list of contaminated waterways extended to 33 states, and the economic impact was quickly felt. When the Atlanta regional office of the Federal Water Quality Administration (FWQA) announced that fish in the Mobile and Tombigbee Rivers exceeded the 0.5-ppm action level, Governor Wallace called on the federal government to declare Alabama a disaster area. In addition to the damage to commercial and sport fishing on inland waters, contaminated ocean fish were also discovered as well as other environmental damage from mercury.

Although quick revisions in freshwater sport-fishing restrictions decreased the economic blow from mercury contamination during and since the summer of 1970, a new shock was administered that December when some popular marine fish tested over the FDA 0.5-ppm action guideline. Because the oceans measured only 0.03 ppb of mercury, they seemed an unlikely source of contamination (151). Nonetheless, a chemistry professor at the State University of New York, Dr. Bruce McDuffie, reported that a can of Grand Union tuna he had purchased at a local grocery store contained 0.75 ppm of mercury (152). Since tuna normally feed on the high seas far from shore and human contamination, they were assumed to be virtually contamination free. But McDuffie's results were confirmed when the FDA tested 138 samples of several brands of tuna. The highest was 1.12 ppm, and they averaged 0.37 ppm (153).

The FDA's nationwide sampling indicated that 23% or 207 million of the approximately 900 million cans of tuna packed in 1970 were above the 0.5-ppm mercury action level. The samples represented several species of tuna from both the Atlantic and Pacific Oceans. Therefore, the sources of mercury could not be identified specifically (152). As some canned tuna was recalled, however, estimates of the percentage that was contaminated steadily decreased. At first 921,000 cans of tuna were recalled. FDA Commissioner, Charles C. Edwards, declared that the fish were absolutely safe to eat at the same time that he ordered them withdrawn from the market. And Charles Carry, Director of the Tuna Research Foundation at Terminal Island, California, reassured the public that most of the tuna was recalled before it reached grocery market shelves (153). The reassurance was needed to counteract negative publicity for the tuna industry in southern California, where 6000 cannery workers and 2000 fishermen pack 65% of all tuna consumed in the United States. Of the 1 million tons of tuna caught throughout the world each year, about 400,000 tons are consumed in the United States (154).

By May 1971 further tests showed that only 3.6% of the cans of tuna were over the 0.5-ppm action level (153). The FDA subsequently reported that only 1 to 2% of the samples were excessive. The original projection was lowered, because most of the contaminated fish were of two species, big-eyed tuna or yellow fin tuna. Consequently, approxi- 12.5 million cans were withdrawn altogether instead of the original estimate of over 200 million. After the National Canners' Association conducted extensive tests under FDA supervision, codes were publicized for the brands of tuna that were either recalled or not issued for sale.

Despite the revised estimates of how much was contaminated, one FDA official said that the tuna canning industry lost $84 million (155). Agents of the tuna industry also insisted that they destroyed the recalled tuna and bore the financial loss. However, rumors circulated that some recalled tuna was exported to countries where food regulations are more relaxed. When Bruce McDuffie was asked about this at a Water Quality Symposium in 1972, he replied that such a practice would be "the sort of a reverse lend-lease that I don't approve of" (141). However, federal policy remained cloudy, because FDA officials continued to insist that the fish were safe to eat while they were being recalled. Dr. Albert C. Kolbye, Jr., Deputy Director of the FDA's Bureau of Foods, said a person would have to eat two cans of tuna containing more than 0.5 ppm of mercury every day for a year before he would experience a toxic reaction. Nonetheless, another 843 cases were recalled in 1973 (156).

While the FDA tried to reassure the public and still monitor mercury in tuna fish, Professor McDuffie exploded another bombshell on Christmas Eve 1970. When a fishmonger delivered swordfish by mistake after McDuffie ordered more tuna, he analyzed them anyway and then notified the FDA that Samurai, a brand imported from Japan, registered at least 1.3 ppm of mercury. Once again the FDA confirmed McDuffie's results. Among 62 samples of frozen swordfish steaks, fillets, and chunks, the range was twice that of tuna, from 0.18 to 2.4 ppm with an average of 0.93 ppm.

With a news story on the wire that 89% of all swordfish sold in the United States could exceed the FDA's 0.5-ppm action guideline, that agency impounded 25,000 lb of swordfish at the Holly Seafood Company. The FDA tested up to 1.02 ppm of mercury, but company spokesmen insisted that their private laboratory results were much lower. Altogether, 400,000 lb of swordfish were eventually either seized or voluntarily withdrawn from the market.

Senator Philip Hart, D-Michigan, later contended that the swordfish publicity brought to light "the first case of human illness in this nation directly attributable to mercury poisoning from ordinary marketable food." Until mercury poisoning was suspected, the victim, a 44-year-old Long Island housewife, had received psychiatric council for 2½ years for what had been diagnosed as a psychosomatic complaint. After she had eaten swordfish daily for 9 months during 1964 and 1965 and lost 45 lb on a Weight Watchers[R] program, such symptoms as lethargy, frequent headaches, blurred vision, and trembling hands were attributed to weakness from the diet. But the condition became more serious each time she resumed the diet for a month 2 or 3 times a year until November 1970. Her dizziness, tremors, and sensitivity to light were gradually accompanied by speech and hearing problems as well as a loss of motor coordination that gave her difficulty deciding which foot to put before the other when walking. She sought medical attention in May 1966 when her memory had become so poor that she could no longer remember the names of her three children, street addresses, or familiar telephone numbers.

No physical cause was indicated until 1971, when the housewife was first examined for mercury poisoning. Although she had not eaten swordfish for 5 months, her hair registered 42 ppm of mercury compared with a national average of 2.0 ppm, and her blood levels were 6 times normal. However, experts disagreed on whether her condition could have been caused by alkylmercury poisoning, since many of her symptoms had disappeared, and these compounds cause permanent neurological injuries (157). However, compensatory brain cells have often permitted some recovery of previous functions among documented alkylmercury poisoning victims.

Two other dieters similarly had unconfirmed symptoms, because their blood and hair levels were analyzed after they stopped eating contaminated fish and their symptoms had disappeared. After eating an 8-oz portion of swordfish four or five times a week from May 1969 until October 1970, a 53-year-old insurance administrator suffered insomnia, loss of alertness, fine tremors, anxiety, and a feeling that he was "coming down with something." The symptoms disappeared after his doctor recommended that he vary his diet, but his blood still measured 18 ppb of mercury. In a similar progression a 26-year-old woman lost her taste for swordfish and then no longer had numb fingertips or trouble focusing her eyes. Her blood measured 21 ppb of mercury (158). Other diet club members who consumed more than average portions of fish measured 60 ppb of mercury in blood samples. Since swordfish typically contain 1.0 ppm of mercury, one 7-oz steak eaten each week could induce blood levels of 20 ppb in a 150-lb human being, and several fish meals per week could insure that more mercury would be concentrated than eliminated.

The swordfish industry was virtually destroyed in May 1971 when the FDA advised Americans to stop eating this fish altogether. Tests on 853 samples revealed that all but 42 exceeded the 0.5-ppm guidelines (159). Before that, a California restaurateur, Todd Ghio, noted that orders for swordfish dropped from an average of 500 to 30 a day when the tuna fish scare began and subsequently rose or fell with the newspaper headlines

(160). Adverse publicity about one type of fish caused the public to shy away from others as well. However, fish speciality shops reported a range of reactions. When the last Chappy's Authentic English Fish and Chips Shop closed in Winnipeg, Canada, the franchise owner blamed mercury pollution as well as a lack of customer acceptance. On the other hand, operators of the H. Salt Esquire shops thought their business was not affected, because the fish came from salt water around Iceland rather than from freshwater sources. They did not serve tuna or swordfish.

The puzzle over mercury in marine fish was increased when museum specimens of seven tuna caught between 52 and 93 years ago and a swordfish taken 25 years ago tested at levels equivalent with contemporary specimens (151). A skipjack tuna captured off the coast of Massachusetts in 1878 contained 0.64 ppm, and among skipjack, albacore, and bluefin tuna collected between 1878 and 1909, the mercury ranged from 0.27 to 0.64 ppm. Samples of albacore and skipjack tuna taken in 1971 ranged from 0.31 to 0.38 ppm. External factors like mercury in the preservative make the accuracy of data on museum specimens questionable. However, these levels are reasonable if estimates are accurate that the mercury content of oceans has increased less than 1% over the years. In this case human beings may long have consumed mercury in marine fish without ill effects or without proper diagnosis.

Another possibility is that numerous contaminants nullify each other. Japanese quail that were experimentally fed methylmercury in tuna fish did not develop toxic symptoms as readily as quail fed corn soya containing methylmercury. Since tuna has a relatively high content of selenium, this element may inhibit the absorption of methylmercury. It is not known if the selenium content of the oceans has risen at an equivalent pace with the mercury levels. Both need better monitoring to determine their impact on human beings (26).

FISH CONTAMINATION, OTHER LOSSES

Fish Protein Concentrate

Even before mercury contamination halted commercial fishing, over the years the Great Lakes Fisheries have gradually altered until lower quality game fish and bottom-feeding scavengers have generally replaced higher quality game fish. In 1920 the commercial catch from Lake Erie was dominated by five high-value fish: cisco, blue pike, sauger, whitefish, and walleye. By 1969 only the walleye remained, and its numbers were in a serious decline. Excessive fishing, temperature changes, and rapid environmental degradation have been the major contributing factors (161). Although sudden influxes of toxic contaminants have caused an increasing number of fish kills in many waterways, the gradual elevation of mercury levels may prevent even lower-grade fish from being utilized as a source of food.

In Norway a rich source of animal feed is fish meal made from herring, mackerel, capelin, and Norway pout (162). The fish are ground up and the liquid is pressed out. Then they are dried into cakes that are 80% protein. But drying also concentrates the mercury. Thus levels are already so high in fish from many American waterways that fish meal must be excluded from human diets. Alewives from Lake Michigan have concentrated 0.90 ppm. Similarly, Gulf menhaden from the Mississippi coast had 0.51 ppm, and Atlantic herring off the Massachusetts coast averaged 0.60 ppm (163). Carp from eastern Lake Erie had 0.58 ppm, which concentrated to 3.12 ppm as fish meal. In general, the protein concentrate is 5 times higher in mercury than the fresh fish. Thus if the whole fish had 0.25 ppm, an average person would consume the maximum allowable daily intake (ADI) of 0.030 mg in only ⅔ oz of fish protein concentrate. A person could consume only ⅓ oz of concentrate if the fresh fish had 0.5 ppm (164).

Whether fish protein concentrate can still be fed to animals will depend on the different species' mercury tolerance levels and whether they are part of the human food chain. Since cats have consumed enough mercury from whole fish to be poisoned, concentrated fish protein would be a dietary hazard for them. A resident of Washington, D.C. reported to the Federal Water Quality Administration that his cat showed symptoms of mercury poisoning after an exclusive diet of smelt. The cat's liver and kidneys had mercury residues comparable with poisoned Japanese cats (165). However, rats fed a diet of fish protein with from 0.17 to 0.28 ppm methylmercury were able to grow and reproduce normally (166).

If mercury levels continue to be monitored and fish rejected for human consumption, this may bode ill for domestic pets. In 1971 the FDA released 2 million lb of sperm whale meat to the pet food industry with the stipulation that it not be used for human beings or cats because of their unusual sensitivity to mercury. Furthermore, not more than 5% or a total of 0.1 ppm of mercury was to be included in the final product that would be monitored in private laboratories and the results submitted to the Bureau of Veterinary Medicine for review. This concession ended a decade in which animals were protected from contaminated foods, a policy that had been implemented at the request of veterinarians (167). The lowered standards for pets also have implications for poverty-stricken citizens who also have acknowledged that they eat pet food.

Fish Kills

Although the gradual accumulation of mercury may harm those who eat the fish, a sudden influx of toxic contaminants is more apt to kill the fish themselves. Many kills result from a complex interaction of numerous pollutants to which mercury may be only one contributing cause. Since 1960 records have been kept on the increasing numbers of fish kills by municipal and industrial waste contamination. The largest single fish kill occurred at Plant City, Florida, in 1969. After untreated industrial and municipal effluent had been dumped into Lake Thonotosassa for 15 years, the organic pollutants decreased the dissolved oxygen concentration in the water until the fish could no longer survive. Although both municipal and industrial pollutants often contribute to the degradation of the waterways, sometimes one or the other can be identified as the specific cause of a fish kill. In 1969 municipal waste caused 1.2 million fish to be killed, whereas industrial waste destroyed an estimated 28.9 million more (168). In 1970 the fish kills in the United States increased 36%. In 634 separate incidents 23 million fish were killed in 45 states. Five of these kills each exceeded 1 million fish (169). Over 73 million fish were killed in such incidents in 1971, but the Environmental Protection Agency reported that the number decreased to 18 million in 1972 (170).

One fish kill that was traced to mercury demonstrates the casual manner with which some toxic chemicals have been handled and the difficulty of pinpointing the cause of such disasters as well as the potential hazard to human health. When fish began dying in the Watauga arm of Boone Reservoir, Tennessee, in the summer of 1968, a biologist and a bacteriologist from the Water Quality Branch of the Tennessee Valley Authority hastened to the scene because it endangered several municipal water supplies drawn from the reservoir (see Figure 10). However, this fear decreased when fish did not die in tanks of water that had been filtered with processes similar to those used to finish drinking water. But the fish died within 2 hours as the reddish-brown coloring spread down the river.

For 5 days the contaminant remained in the upper layer of the stratified reservoir without spreading into the colder water below. By the time a rainstorm diluted the poison, an estimated 500,000 fish were dead. Most of them were gizzard shad, carp, catfish, and sunfish. The poison caused them to swim spasmodically near the surface and blow air bubbles. They seemed to hemorrhage about the dorsal fin as they lost sensitivity to the

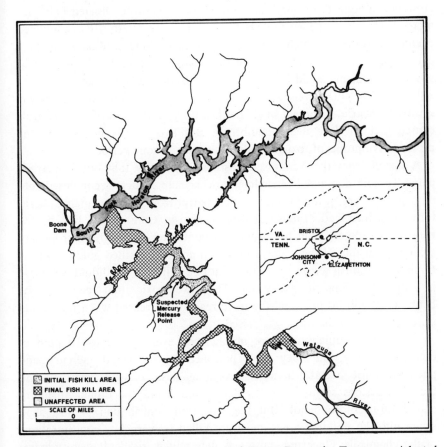

Figure 10. Fish kill in the Watauga arm of Boone Reservoir, Tennessee. Adapted from *Fish-kill in Boone Reservoir,* July 9–13, 1968, Tennessee Valley Authority.

point of being relatively easy to touch. Then they gradually stopped swimming and lost their equilibrium before they died (171).

Although rain ended the fish kill before the cause of poisoning was identified, samples of water collected initially showed 180 ppb of mercury. Near the center of the kill area, diphenylmercury had concentrated to 200 ppb or more than 200 times the amount that would kill half the fish during a 2-hour exposure. Even with the contaminant identified, the cause was not discovered for more than a month. Then investigators returned to sample the contents of 29 of the approximately two hundred 55-gallon steel drums that supported docks or floated loose near the shoreline. Local residents assumed that the barrels were discarded oil drums. However, among those that had washed up on shore or floated half submerged, often penetrated with bullet holes, some 60 to 70% contained chemicals, some of which were extremely toxic. Since ordinary pumping equipment normally did not empty the barrels, several gallons of full-strength chemicals could have been in the drum or drums that caused the kill. Twenty-eight barrels contained one or more gallons of liquid. One syrupy, industrial detergent made the fish tipsy, and seven other chemicals caused them to lose their equilibrium. But two barrels induced all the previous symptoms of toxicity within the same 2-hour period.

A barrel from the Tennessee Eastman Company that bore the trade name Tecsol[R] contained phenylmercuric acetate, which decomposed into diphenylmercury in the water. Another barrel from the Buckman Laboratories was inscribed "Industrial Microorganism Control." It had floated two-thirds submerged in the water and contained a 2,4,6-trichlorophenol solution that turned the water a brownish color and killed goldfish within 6 minutes. Sunfish were so sensitive that one gallon of the liquid was toxic in 230,000 gallons of lake water. The lower toxic limit seemed to be between 20 and 50 ppb, and the toxic reaction increased when the acidity or temperature of the water was raised.

Algae and zooplankton were not harmed even by up to 10 ppm of diphenylmercury, but they were poisoned by elemental mercury and phenylmercuric acetate. Some fish such as bass, bluegills, darters, and black silversides seemed able to detect and avoid the diphenylmercury and 2,4,6-trichlorophenol. But the red-eared sunfish and gizzard shad swam into the polluted areas and were killed by very low concentrations of the chemicals. Once the cause was determined, residents were admonished to be more careful with old barrels to prevent such kills in the future. In these discussions similar previous incidents were also recounted. For instance, in 1964 newspapers had reported that fishermen seined a nearby creek and tossed the easily caught bass of all sizes to friends on shore. When the remaining fish began to die, no impact study was conducted to determine if the fishermen suffered any ill effects from eating the large numbers of fish that had been taken home the preceding week (171).

6

THE SACRIFICIAL LAMMS

Like the Americans the Canadians discovered mercury contamination in such unlikely places as the northern tip of British Columbia. In waterways where king salmon still challenged fishermen, crabs registered up to 6 ppm of mercury in mud flats where waste had seeped out of a holding pond that received 20 lb of mercury per day from the F.M.C. Chemicals Company. Even the Ottawa River in Canada's capital city was contaminated by discharges from several pulp and paper mills upstream until the fish had up to 2 ppm of mercury. The initial impact of mercury contamination on tourism in the north country reads like a case study for *Future Shock* (172).

As the quality of life deteriorates and tensions increase in the polluted American cities, fly-in fishing camps have become increasingly popular with hunters and fishermen who retreat to the Canadian wilderness to pursue wild game, birds, and fish while they forget the smog and hustle of cities like Chicago, Detroit, and New York. In 1947 Barney Lamm and his wife Marion opened the first fly-in fishing camp on the English River chain at Ball Lake in Ontario province. From a few log cabins the business expanded to include a fleet of airplanes to transport guests to all 17 tourist camps on the Wabigoon–English–Winnipeg River system (see Figure 11). Barney's Ball Lake Lodge, the largest of them all, grossed a million dollars in 1969 but did not open at all in 1970. North of Kenora the fly-in fishing camp owners and the Ojibway Indians were suddenly confronted with new environmental, social, political, and economic problems that

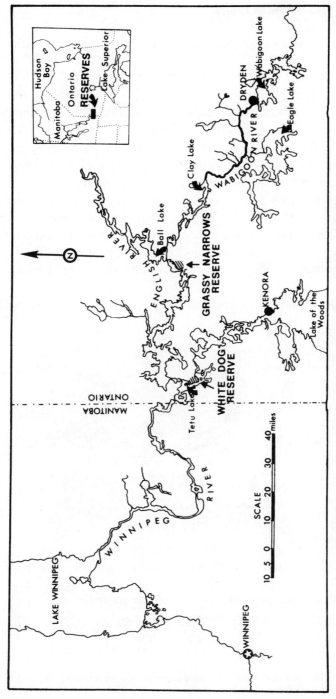

Figure 11. The Wabigoon–English–Winnipeg River system and Indian reserves.

73

ultimately may destroy the last vestige of the Indians' traditional way of life.

On April 23, 1970 the camp owners on the English and Wabigoon River called a meeting after Ontario provincial officials had informed them that fish in Clay Lake contained unacceptably high mercury levels. The owners then agreed that their camps should not open. They had not been given the actual figures, only that the fish exceeded 0.5 ppm. Fishing would still be permitted under the Fish for Fun policy, but warning signs would be posted in conspicuous places. The signs displayed a frying pan with a fish dramatically crossed with a big red X and a hazard warning (see Figure 12). This bad news came after the camp owners had spent the winter promoting their camps at the sport shows in the United States. Larders were being stocked for what promised to be a record season, but negative publicity could undo all the promotion and advertising.

Although fishermen would still be able to take home "one for the wall,"

Figure 12. Canadian health warning sign posted near waters containing fish contaminated with mercury.

a trophy fish to be mounted instead of eaten, they might not be willing to pay as much as $10 per pound to catch fish that they had to throw back. Besides losing the opportunity to "bring back the big ones," the guests would also have to relinquish another pleasure of the northern fishing trip, the shore lunch. Camp owners had long promoted this culinary institution as a traditional speciality of the fishing camps. Only the visiting fishermen could evaluate how much it contributed to their total vacation pleasure. After a cold morning of fishing, the guests were delighted to have the Indian guide pull the boat up on shore where they could warm themselves by a roaring fire while the guide filleted some of the morning's catch, made coffee, and heated up cans of beans and spaghetti. The highlight of the meal was the freshly caught fish fried over an open fire. The guide shared the outdoor picnic with guests who would often eat four or five of the fillets. Now guests were asked to eat the beans with substitutions for the fresh fish.

Fishing was initially banned from Dryden to Clay Lake. A week later Ball Lake was also placed on the contaminated list as it was extended 150 miles farther down the Wabigoon–English River system. Camp owners had to decide quickly whether to adopt the Fish for Fun policy or close, because the approximately 155-day season was scheduled to begin on May 9th. With the first guests expected in 2 weeks, the Lamms had already invested over $70,000 in promotion for the season. To accommodate a guest list of from 70 to 100 people per week, an equivalent number of employees—kitchen help, guides, cleaning women, and waitresses— were already being flown in along with large quantities of food and other supplies. Besides the financial loss, closing could cost the good will of guests who had planned fishing vacations at Ball Lake and would make reservations elsewhere.

The Lamms decided to close for one season. Since they doubted that guests would accept the Fish for Fun policy, they would be moved to other lodges on uncontaminated waterways. To contact them quickly, Barney Lamm announced his decision in newspapers, on a Chicago television station, and in individual letters. An example is reproduced as Figure 13. Relocating the fishermen would retain their confidence, Barney reasoned, so they would return when they were reassured that the mercury was gone. Neither Barney nor the government officials were aware that they were facing a long-range problem with no solution.

However, as Barney Lamm headed for Toronto to find out more about

We regret to announce that due to mercury pollution in some lakes in the territorial waters of the district of Kenora, Barney's Ball Lake Lodge is not open. The lodge is ready to operate but we do not feel we can accept any guests until such time as the mercury pollution in the water and fish is at a safe level for human consumption. These levels are set by the International Health Authorities.

My business has been built on being honest and frank with my guests. I have been advised by the authorities that at the present time, the fish from these waters should not be eaten. Therefore, under the present circumstances, I do not feel that you could have a safe and enjoyable trip. The source of the pollution has been stopped and I am hoping that it won't be long before all of our valued guests will be able to return for another memorable fishing trip to Barney's.

We will be taking tests of the water and fish frequently and will keep you informed of the situation during the summer. Until such time as Barney's Ball Lake Lodge will be able to reopen, let us help you arrange a fishing trip to a lodge located on a lake that does not have mercury problems.

For further information write: Barney's Ball Lake Lodge, Box 50, Kenora, Ontario, Canada or phone, Area code 807 468-6092 or 468-9121.

BARNEY LAMM
BARNEY'S BALL LAKE LODGE.

Figure 13. Bulletin from Barney's Ball Lake Lodge warning tourists about the mercury-contaminated fish.

the mercury-contamination problem, on September 13, 1970 on a Kenora radio program, George Kerr, Minister of Water Resources, announced that the problem would last no longer than 16 to 20 weeks. He said dredging would begin on October 1. Although Minister Kerr was not sure where the dredging wastes would be deposited, he expressed hope that the costs would be shared by the polluting industries and that the lakes on the Wabigoon-English River system would be free of contamination by the following spring.

Although mercury contamination was a new threat to tourism in the Kenora area, pollution was not. In fact, the president of the Patricia Regional Tourist Council, E. L. Palmer, contended that publicity about mercury would be superfluous because "no one in his right mind had fished the Wabigoon River in the Dryden area for fifty to sixty years" (173). The Dryden Paper Company Limited had dumped pulp wastes into the Wabigoon River until nothing could live in it for 30 miles. Solid waste particles half filled the water, and some people complained that the fish had an unusual taste even in Clay Lake 50 miles downstream. The new threat, mercury, came from the Dryden Chemical Company, an affiliate of the paper mill. The chloralkali plant had begun production in 1962. Each year thereafter the mercury cells producing chlorine and caustic soda were replenished with from 700 to 12,000 lb of the heavy metal.

A ministerial order prohibited the firm from dumping any more mercury into the river after May 1, 1970, and Dryden Chemical Limited subsequently announced that their mercury losses were reduced from 1.0 to 0.5 lb per ton of chlorine. In May Leo Bernier, member of the Ontario Provincial Parliament, later appointed Minister of Mines and Northern Affairs, triumphantly announced that Dryden Chemical had beaten its pollution problem. The company had budgeted $2 million for antipollution facilities to be installed during the next 2 years, beginning with a settling tank to catch impurities before they could reach the river (174). On July 23, 1970 the *Toronto Daily Star* quoted the chief engineer, Bernie Ford, as saying the mercury spills had been stopped completely (175). Nonetheless, the firm continued to manufacture liquid chlorine and caustic soda with mercury cells and to replace as much mercury annually as before (176). In 1973 the Canadian chloralkali industry reported that settling ponds at many such plants were overflowing and that half the mercury that was used remained unaccounted for (177). Whether or not Dryden Chemical Limited had stopped discharging mercury, residues accumulated over the previous 10 years still remained in the river to be methylated.

Since mercury contamination could not be seen or tasted and no one had apparently suffered any ill effects, it was easier to blame bad publicity when camp owners saw what promised to be one of their best years turn into one of their worst. After Barney's Lodge closed, business was off from

25 to 60% even at Hook's Separation Camp, the second largest camp in the area. When guests could not eat the fish that they caught locally, they were also suspicious of fish that were flown in for the shore lunch. The whole Kenora business community suffered. Unemployment rose because the camps had accounted for 10% of the local jobs. In addition, fewer fishermen stopped in Kenora to buy licenses and supplies from local store-keepers, and sportsmen could no longer augment their own catches by buying fish from commercial fishermen. The Chamber of Commerce re-ported that the area suffered a 25% economic loss, from $6 to $10 million, because of mercury pollution. Vacationers spent an estimated $6 million less than the year before, according to calculations based on the number of U.S. dollars cleared through the chartered banks in Kenora where most of the local tourist camps did their business (178).

Barney Lamm shared the blame for the bad publicity with provincial government officials who ordered the warning signs posted prominently near the camps at the beginning of the season. Later, the signs were rele-gated to obscure locations, and many disappeared entirely. Some lodge owners were rumored to be paying the Indians a bottle of whiskey for each sign they brought in (179). In light of all the publicity about mercury in the spring of 1970, it is impossible to determine how much Barney Lamm's decision to close affected business at the other camps. Since his camp was the largest on the English River system, the closing did make headlines in major cities throughout the United States and Canada. Subsequently, local tourist trade journals combined optimistic reports that mercury pollution was overstated with vehement attacks on Barney Lamm. According to one journal the disastrous 1970 season was caused by a "black sheep in the family who fed the newspapers, radio, and television in the United States a much blacker picture of the pollution problem than actually existed" (180). As the publicity decreased and the tourists returned, it became easier for local people to believe the mercury contamination had all been Barney Lamm's bad dream.

But the nightmare was real for the Lamms, who set out to learn more about mercury contamination after Ball Lake Lodge was closed. When she compared warnings on both sides of Lake St. Clair, Marion Lamm found that fishing licenses purchased on the Canadian side had no notice of the mercury hazard, whereas the Americans issued a warning not to eat the fish when a fishing license was purchased. Despite poor business on the Canadian side, small camp operators were still open. Many of them had accepted loans from the provincial government with the stipulation that they remain open to be eligible for the money. And the name of Norvald Fimreite was either praised or damned around Lake St. Clair, but little information on mercury was available.

Therefore, Mrs. Lamm went to Western Ontario University in London, Ontario, where Fimreite was preparing to defend his doctoral dissertation before returning to Norway. He agreed to survey mercury levels in fish in Ball Lake and surrounding waters, thus providing information the Lamms had not been able to obtain. Fimreite got a collectors' permit that allowed him to "kill, capture, and possess for scientific purposes mammals and birds, as well as fish." For 3 weeks he collected, weighed, and measured a wide range of specimens before the District Forester notified him that the Ontario Department of Lands and Forests had revoked his permit. The sampling was ended, but 510 fish, mammals, and bird specimens were sent to a commercial testing laboratory. The results were the highest levels of mercury yet reported for freshwater fish anywhere in the Western Hemisphere and possibly the world. Healthy-looking northern pike tested up to 28 ppm, walleye at 20 ppm, bass at 10 ppm, and burbot at 25 ppm (118, 181). In Lake St. Clair pickerel had tested the highest with almost 7.0 ppm. At Minamata the record was 24.1 ppm in a disabled fish that floated on the surface and 39.0 ppm in shellfish that inhabited the bay. Japanese fish of different species showed signs of illness with as low as 8.72 ppm (28), whereas in Sweden 20 ppm were found in apparently healthy pike (50). The same was true in Ontario.

Until Fimreite completed his study, the Lamms had not known whether raising the accepted tolerance level to 1.0 ppm would be of any help. This had been done in Sweden and was being proposed in Canada. The Lamms paid $55,000 to learn that some fish exceeded the 0.5 ppm standard by 30 or 40 times. When government health authorities were confronted with their findings, the results were unofficially confirmed. About the same time *The Toronto Globe and Mail* reported that burbot taken from Clay Lake had from 10.46 to 16.65 ppm of mercury, 33 times the accepted tolerance level (182). The government's initial findings have never been made public, but other studies confirmed the high levels (118, 181, 183).

When the Lamms attended the International Conference on Mercury Pollution at Ann Arbor, Michigan in the fall of 1970, many experts had not heard about the mercury contamination in the Wabigoon–English River system. Moreover, Swedish and Japanese scientists could offer no assistance for a tourist camp operator who was worried about the Indian guides. A Canadian newspaper reporter, Betty Lou Lee, later described the scene.

> And Barney Lamm stood up in the midst of all the scientists yesterday and asked a Japanese doctor who'd had experience with people who died of mercury poison, just what the danger was to people eating these levels of fish.
>
> There were scientists who looked around askance and a bit annoyed when Barney Lamm said he was a tourist camp operator.
>
> One of them even questioned that he knew what he was talking about when he gave levels that high. But a representative of the Fisheries Research Board in Winnipeg confirmed that Barney Lamm knew what he was talking about because the fish were tested by the fisheries board and that's what they found (184).

The Japanese delegate, Dr. Tadao Takeuchi, was sure that the mercury levels in Ball Lake were dangerous, but he could not say how many fish had to be eaten before a person's health would suffer. The Lamms left the conference convinced that mercury pollution was a long-term problem that could destroy the Indians' health and way of life. The camp owners could relocate, but the Indians were trapped.

The Ojibway Indians from White Dog and Grassy Narrows reserves had adapted their ancient customs to the advent of modern civilization by working as guides for the fishermen who came to the lodges for 5 or 6 months each summer. The guides led fishermen to hidden spots where the fish lurked, and they cooked the shore lunches while their wives worked in the camp kitchens or cleaned the cabins. Some Indians had also built commercial fishing businesses on the Wabigoon–English River system. In the winter they returned to their reserves and eked out a relatively independent existence by setting trap lines, although the numbers of fur-bearing animals such as mink and otter had also been declining in recent years (181).

Barney Lamm had employed up to 60 guides each summer. They lived with their families in a small village of log cabins not far from Ball Lake Lodge. In all, the fishing camps employed approximately 225 Indian guides. The fishing ban cost them and commercial fishermen an estimated annual lost income of $380,000.

Whereas the government promised a loan of up to 70% to compensate the commercial fishermen, the guides were left with one cash crop, the wild rice harvested in the fall, and welfare—the White Man's way. John Henry, a member of one Ojibway tribe, described how closing the Ball Lake Lodge affected Alec Necanapenace and his family. When he lost his job as guide, Necanapenace and his wife Mathilda had to raise their six children on $240 a month in welfare. Like the other 500 Indians who live on the Grassy Narrows Reserve, the Necanapenace family had to eat fish or starve when their welfare money ran out (185). The English River was still the most accessible waterway, although the Indians were warned not to eat the fish. In 1975 the new chief, Roy McDonald, pointed out that his people could not afford gasoline to drive to less contaminated waterways. Nor were all the Indians convinced that it was harmful to eat local fish, especially after they were rehired as guides when the tourists returned. If they ate fish during the summer without ill effects, they saw no reason not to do the same in the winter. Nor was contradictory evidence provided. Government health officials tested the Indians for mercury in 1970 but did not give them the results for 3 years. The challenge of "Show me someone who had died of mercury poisoning" also went unanswered, because health officials conducted no autopsies on Indians. At Ann Arbor Dr. Takeuchi offered to conduct pathological studies of Indian brain tissue, but none was made available to him (186). Dr. Masazumi Harada made a similar offer in April 1975 and bluntly assured a Canadian federal mercury pollution board that the results could and would confirm Minamata disease among the Indians.

Without direct evidence of poisoning, the Indians' culture made it difficult to determine if changes in their emotional and physical health were induced by mercury or other factors despite apparent symptoms. For example, Stanley Indian demonstrated a marked psychological change. He was a popular guide at Colin Myles' camp for many years, and guests requested his services each year when they returned to the lodge, but he gradually became so withdrawn and morose that he had to stop guiding. Even after he died, however, no tests were conducted to determine if mercury poisoning had occurred. Another Indian guide, Art Anderson, went blind and into a coma before he died. Again the cause was not determined.

The tremors exhibited by many Indians on the streets of Kenora are usually attributed to alcohol, and the Indians acknowledge that many of their numbers turned to this and other drugs after their lives were disrupted when their reserve was moved. Such excesses may also cause the many birth defects that could result from inbreeding or chromosome damage from alkylmercury poisoning. Although industrial workers become shy and suspicious as symptoms of mercury poisoning, the Indians have always felt unwelcome among the townspeople. Since many of them have already lost their jobs, their fears may be legitimate as well as symptomatic.

One member of the Ojibway band, Thomas Strong, was finally autopsied in August 1972 after he died of a heart attack at age 42. The local coroner did not conduct the inquest in Kenora. Instead, the Indians contacted Dr. H. B. Cotnam, Chief Coroner of Ontario (187). Strong's mercury levels were so high that Cotnam decided the blood samples must have been contaminated. At the Indians' insistence the Mercury Task Force of the Ontario Ministry of the Environment released the 1970 blood and hair test results in March 1973. Members of the Ojibway tribe at Grassy Narrows and the Walpole tribe on Lake St. Clair had 12 times the normal levels, up to 96 ppm in their hair. Japanese at Minamata had begun to show symptoms of poisoning with 145 ppm in their hair, and they were severely affected at between 500 and 600 ppm (188). With from 40 to 150 times more mercury in their blood than the average Canadian, some Indians exceeded the levels at which neurological damage is known to begin. Dr. G. J. Stopps of the Ontario Provincial Health Department said it was a wonder more Ojibways were not dying, because some of them had blood mercury levels that exceeded those of fishermen who died at Minamata. The Ojibway Indians had continued to eat fish daily at least during the tourist season and to accumulate mercury for 3 years after they were tested, although the Walpole Indians reportedly ate only an average of 10 lb of fish per year (189). When the tests were finally released, newspapers reported that unidentified health officials had called withholding the Indians' blood tests from them a form of "genocide by neglect" (190). Keeping the 1200 Ojibway Indians ignorant at Grassy Narrows and White Dog reserves was labeled a "modern-day horror story."

Although the Indians are a small minority group that did not arouse great public concern over their plight, the possible effect on other segments of the society is also being ignored. Not all the guides are Indians, and at least one other local man developed such difficulty walking and talking that he could no longer guide. Moreover, many local residents maintain private vacation camps and consume more fish than the national average. Not only has Kenora, a town of 11,000 population, had to expand the jail to house more drunk and disorderly Indians, but the schools for the mentally retarded have also been expanded faster than in the rest of Ontario. With the new addition for retarded preschool children (191), facilities are available for all ages. Moreover, the province averaged 15.8 infant deaths per 1000 live births due to pneumonia and related lung disease, whereas Kenora averaged 30.5 from such ailments. Similarly, Ontario province had 20 motor vehicle accident deaths per 100,000 population, whereas Kenora had a third more, 27.5 per 100,000. The suicide rate was also higher with 11.9 per 100,000 population in the province and 16.6 in Kenora. Whether the Indian population accounted for the higher percentage was not determined (192).

When the first Fish for Fun season ended in the fall of 1970, losses had consumed the profits of the previous year, and the Kenora camp owners, now ironically renamed the Fish for Fun Camp Owners Association, found themselves entangled in a web of federal, provincial, and local politics. Government officials no longer talked of dredging the waterway. Instead, Canadian Fisheries Minister Jack Davies had announced that "we have caught our mercury problem in time before there was any real danger to human beings" (193). Despite this assurance, camp owners on the brink of bankruptcy asked government officials for assistance to promote the 1971 season at the winter sport shows in the United States. They faced stiff competition from camps advertising that the fish were "clean" in their waterways.

When John R. Roberts, federal member of parliament, met with camp operators, they suggested several alternatives by which the government could minimize the losses from mercury contamination. Their preference was to declare the fish safe to eat or at least to clarify which areas were contaminated so the tourists could be certain. This had not been done in 1970. While transferring guests from Ball Lake to lodges on the Winnipeg River, Barney Lamm inadvertently learned from a commercial fisherman that this waterway was also contaminated. The information had been given to the Fresh Water Fish Marketing Corporation, which was responsible for testing fish from contaminated areas. The Ontario Ministry of Lands and Forests had also published a bulletin listing lakes and the species of fish over 0.5 ppm of mercury, but the bulletin was not widely circulated, and the 1971 edition deleted the notation that some fish in polluted Ontario lakes contained up to 15 ppm of mercury. The new edition only listed the species of fish over 0.5 ppm in each lake (194).

Camp owners on contaminated waterways requested that the government purchase their camps so they could relocate elsewhere, and federal minister John Reid also recommended that the Canadian government compensate the camp owners. He cited a precedent set 15 years earlier when a new mine at Steep Rock Lake, Ontario, had destroyed the businesses of commercial fishermen and tourist camp owners on the Sein Ruin River system. The owners were compensated and then allowed to buy their property back for $1.00 if they did not want to relocate (195). But the federal government decided mercury pollution was a provincial matter on the Wabigoon–English River system which is entirely in Ontario province, although it flows into the Winnipeg River system in Manitoba province. Ontario provincial officials recommended that small camp operators sue the polluters, a course that Reid described as highly impractical, because they lacked the finances to compete with large industries. Barney Lamm is suing the Dryden Chemical Company, but the legal procedure is long and involved, because few precedents have been set.

Camp owners also requested compensation for their 1970 losses, since they had stayed open for the season at the government's recommendation (193). But provincial officials were reluctant to begin an expensive compensation program that would also set a precedent for future pollution cases. Instead, after telling the camp owners that he would take the matter under advisement, Prime Minister Roberts resigned, and the new provincial government offered the camp owners loans to promote their camps at the sport shows in the United States and to winterize the lodges. Lodge owners who accepted these loans optimistically heralded the 1971 season by downplaying the hazard of mercury contamination. Future guests were told that the adverse publicity had been greatly overplayed in the United States. Warning signs were equated with those on cigarette packages, and guests were encouraged to bring their catches back to the camp operators who were still "eating and enjoying" the fish. The 1973 Ontario Mercury Task Force Report recounted that the tourist camps had built their businesses back to the 1969 level "largely by promoting the fact that the mercury problem had not presented a recognized health hazard for their guests or their employees." Trying to stay in business, then, almost inevitably led to downplaying the seriousness of mercury contamination. The mercury remained, but the publicity died out as the issue was shifted from health to public relations.

In 1971 the Ontario Department of Lands and Forests *Newsletter* reported that mercury levels were less than 0.5 ppm in fish from such Ontario lakes as Lake of the Woods, Lake Nipissing, and Lake Simcoe (196). Since the Saskatchewan, Winnipeg, and Wabigoon–English River systems remained highly contaminated, provincial health official Dr. R. B. Sutherland warned that the fish should not be eaten and warning signs should be posted (197). After Sutherland retired in 1971, his successor, Dr. G. J. Stopps, also declared that "there isn't an expert in the world who would say those fish are fit to eat." Stopps contended that eating one 7-oz fish fillet from the Wabigoon–English River system per day for 3 weeks would increase the mercury blood levels to the range of 2000 ppb at which some persons were poisoned at Minamata. Eating only 7 oz of fish per day for a 2-week vacation could leave the mercury blood levels elevated 20% a year later. Thus methylmercury could accumulate from one year to the next, even if none were added from fish or other exposures between fishing trips (196). But these medical cautions were generally ignored.

When no more warning signs were posted either, some resort owners also circulated rumors that Barney's Ball Lake Lodge had really closed for some mysterious, unexplained reason that had nothing to do with mercury. By the spring of 1973 the Lamms were flying fishermen to God's Lake, where the mercury levels were low and local Chippewa Indians acted as guides. Back at Ball Lake Lodge the equipment was falling into disrepair and the log cabins were weathering to an ominous black. At the other camps the Indians were again guiding guests and sharing the shore lunch. Provincial authorities had tested some Indians for mercury at the end of the 1972 season, but many were at the Indian hospital in Sioux Lookout, Ontario, far north of the contaminated English River system. These Indians had low blood mercury levels. The Mercury Task Force annual report again acknowledged that the fish in the Wabigoon–English River system exceeded the 0.5-ppm mercury limit by as much as 30 fold and should not be used for human or animal food. However, when CBS television interviewed American fishermen in Kenora at the beginning of the 1973 season, they said they had not heard about the mercury pollution, but the fishing was great and they certainly intended to eat the fish. On the same program Cabinet Minister Leo Bernier again said he still was not sure that the mercury threat was serious, whereas the Indians said they were told to eat the shore lunch and not mention mercury to the guests or they would be fired. Thus it was essentially business as usual at the tourist camps on the English–Wabigoon River system in 1973 and afterwards.

This lessened the likelihood of government action, although the Provincial Mercury Task Force still recommended that the tourist camp operators be compensated to relocate and the Indians be given an alternate way of earning a living. Welfare costs had increased from $2000 before 1970 to $90,000 in 1973 (178). The Task Force proposed that the Indians be encouraged to try mink farming, lumbering, or commercial fishing in uncontaminated waters before welfare became a way of life. The Indians were given adult education classes at the Grassy Narrows Reserve so they would understand the mercury hazard. However, Chief Assin of Grassy Narrows acknowledged that the Indians still ate contaminated fish when their welfare money was depleted, because "some of my people know too much about being hungry and not very much about this mercury. What else are they going to do?" (190).

7

MUTUAL SUPPORT

THE WEAK STRUGGLE AGAINST THE ODDS

The Indians at Kenora, the poor fishermen at Minamata, and the illiterate farmers in Iraq are all victims of methylmercury poisoning, and all are essentially powerless against the combined opposition of industry, government, and influential citizens who benefit from modern society. It is too late to prevent the contamination that has already led to the destruction of many lives and much of the environment. Concerted moral pressure even by a fairly large segment of the population is most likely to result only in money compensation for the victims. At the 1972 United Nations Environmental Conference in Stockholm, 17-year-old Shinobu Sakamoto haltingly reminded the participants that money would not change the imperfections her body suffered because her mother had eaten mercury contaminated fish before she was born.

> Even when the trial ends, nothing is changed, because my body will not become well again. Even if we get the money, it's the same thing. (198)

Nor were any victims present to describe the major epidemic then in progress in Iraq. The news had been suppressed just as other epidemics were not publicized. Similarly, in Ghana more than 144 persons were poisoned and 20 died from eating mercury-contaminated corn in 1967, but the epidemic was not reported in a scientific journal until 1974 (199). What occurred in one country did not raise moral consciousness in others. Although the wire services flash news stories around the world in minutes, pollution is often treated as a local issue or may be in the headlines one day, after which the public assumes that it is cured. For example, the cause of Minamata disease had been discovered and obscured before the Dryden Chemical Company began to contaminate the English–Wabigoon River chain in Ontario.

Professor Jun Ui of the University of Tokyo contends that the discovery of the hazards of contamination cannot prevent future disasters because the instinct for self-preservation is inherent in bureaucracies as well as individuals. Consequently, the powerful industries continue to contaminate the environment while the weak and poor suffer because they are abandoned. The national attitude toward environmental pollution depends on the strength of basic human rights in that country, and these depend on the will of the citizenry. Nor can science be regarded as the solution to problems that technology has wrought. The experts who conduct expensive and tedious analytical investigations find their results compromised, because their answers cannot be definitive. The scientific method, by definition, is inevitably partial and incomplete (200). Instead, local citizens instinctively grasp the problem and insist on a resolution while scientists are still trying to ensure that their findings are accurate.

The courts also have encouraged procrastination in the handling of pollution cases, in some instances because the judges simply are too far removed from the victims to grasp the implications of their suffering. Before the victims received a favorable ruling against the Chisso Company, they had to pit themselves against the most powerful industrial and political forces in the country. Their strength might well have been insufficient without the student demonstrators who came to their aid. Ui hailed this as the beginning of a direct democracy that "resembles the early Christianity, to build a society on the basis of the unconditional compassion and service to the weak. . . ." Therefore, he suggested that the history of Minamata disease could teach us an important spiritual lesson for the future (200). The battle between the weak victims and strong industry and government has now been waged for 20 years at Minamata.

The faculty of the medical school at Kumamoto University were supported in their research into the cause of Minamata disease by the Japanese Ministry of Health and Welfare, the Ministry of Education, and the U.S.

National Institutes of Health. But their progress was impeded even as people died of acute poisoning. As family members initially hid the victims and the Shin Nihon Chisso Company kept secret Dr. Hajime Hosokawa's discovery that factory wastes could cause Minamata disease, factory workers also refused to aid the fishermen, although they became aware of what the company was trying to conceal at an early stage (201). In 1967 their union magazine finally published an article of self-admonition because of their earlier inaction.

As factory executives denied the Kumamoto researchers admittance to company property to examine the waste discharge, they also refused to disclose the nature of their products or any results of their research that might damage the company's interests. Consequently, the Kumamoto research group wasted time attempting to determine if the inorganic mercury compounds were biologically methylated.

Moreover, in 1959 and 1960 the factory also published several contradictory reports by well-known scientists. R. Kiyoura stated that mercury pollution in Minamata Bay was slight compared with several other districts of Japan, but he did not disclose the locations (202–204). Later, high mercury levels were also discovered where other acetaldehyde manufacturing operations discharged untreated waste into rivers. Not revealing this information delayed discovering the cause of Minamata disease, and Ui has charged that earlier revelations might have prevented the Niigata disaster in 1964. Thus he contends that such secrecy could be considered a criminal act (29).

Some researchers employed by the government also opposed the theory that mercury caused Minamata disease. The Ministry of International Trade and Industry employed a scientist from the Tokyo Institute of Technology who contended that methylmercury compounds did not cause the malady, whereas other scientists announced that the symptoms of Minamata disease could be induced by feeding putrid fish to cats. Opponents of the alkylmercury poisoning theory proposed that a new organization be created to investigate the real cause of Minamata disease, and it was later learned that they were receiving financial support from the chemical industry to oppose the mercury theory (29).

Ostensibly because no conclusion could be drawn from the differing viewpoints, the interministerial ad hoc committee on Minamata disease decided to discontinue any further direct investigation. Consequently, the Kumamoto University research group lost a substantial share of their financial support in 1959, but they continued the investigation. Opposition viewpoints also delayed establishing blame and compensating the victims for several years. Since agreement had to be reached in at least two successive committees, one in the Ministry of Welfare and another in the Board of Science and Technology, this procedure also took much time and decreased the likelihood that the government would support any recommendations that were unfavorable to industry. Thus on September 26, 1968, officials announced that it was impossible to determine if the contamination was a long-range issue or two or more short-term problems. Then the possibility was again raised that the Niigata epidemic was caused by agricultural chemicals that had been stored near Niigata Harbor until the earthquake released them into the waterway on June 16, 1964. With these conflicting opinions the government proclaimed that it was impossible to assess the effects of short-term pollution because of a lack of information.

As industry and the national government delayed and obscured research into the cause of Minamata disease, local political leaders in Minamata city also tried to sidetrack the investigation. The Chisso Company pays about half the city's income from taxes, and the victims were thwarted on all fronts in their efforts to receive more than minimal compensation. In 1960 the Shin Nihon Chisso Company paid condolence money up to 300,000 yen ($833.00) for each of the deceased and 100,000 yen for each adult patient. But factory executives denied any connection between the factory drainage and Minamata disease and insisted that a release clause accompany the benefit payments. The document stated that "even in the event that the company's factory drainage should in the future be found to have causal connection with Minamata Disease, the Mutual Aid Society of Patients' Families shall not make any new demand for compensation" (205). Even in 1968 when the cause-and-effect relationship of the disease and the factory waste was finally acknowledged, the Japanese Economic Planning Agency defended having discontinued the investigation in 1960 because the victims had been compensated (29).

At Niigata the victims became so impatient with the government's inaction that they started a civil law suit against the Kanose factory to obtain compensation for their damages. Begun in June 1967, this was the first pollution trial in the history of the Japanese court system. In June 1969 Mr. Eizo Watanabe also began a new compensation civil lawsuit against Chisso Company in the Kumamoto District Court by saying, "Since today we are forced to fight against the power of the nation. . . ." This imbalanced struggle continues, although in July 1969 the ailing Dr. Hajime Hosokawa finally testified in court that the famous Cat 400 had confirmed the cause of Minamata disease at the Chisso factory more than a decade earlier. Out of loyalty to the company he worked for, the conscience-stricken Hosokawa had kept the results secret as the list of victims lengthened over the years (206).

Their struggle was also carried to the stockholders' meetings of the Chisso Company. After appeals to the public, "indictment groups" were organized by citizens in a dozen cities. Subsequently, in 1970 the Victims' Relief Act passed the Japanese Diet and exempted medical costs for officially recognized victims of pollution disease. Again this was inadequate.

The Chisso Company was ordered to compensate the dependents of the deceased as well as to offer pensions for the permanently injured and restitution for the fishermen's lost pay. This compensation was very limited and difficult to obtain (31). In May 1970 negotiated settlements with the Minamata disease victims ranged from 2.95 million yen ($8194) to 4 million yen ($11,011) for the deceased. Health and Welfare Minister Tsuneo Uchida said this settlement was as high as possible considering that no laws had established liability to indict enterprises for polluting the environment. But *Asahi*, a major Japanese newspaper, editorialized that the small compensation only showed how cheaply human life was valued in Japan when the economy was enjoying a high rate of growth (205).

Although some victims of Minamata disease were finally recognized and the Victims' Relief Act was passed in 1970, new victims went unrecognized and uncompensated. It was politically advantageous to contend that the disease had essentially been identified and halted by 1960. Therefore, no more health inspections were conducted at Minamata, and medical experts in the local prefecture rejected new applications by victims who wanted official recognition to apply for compensation. One of them, Mr. Kawamoto, identified eight others, and the group applied pressure until the federal Environmental Agency also recognized their claims in 1970. Then new health inspections were started in Kumamoto and Kagoshima prefectures, and more new victims applied for recognition: 57 in 1971, 163 in 1972, and 358 in 1973.

The ones recognized as bona fide pollution victims received no answer from the Chisso Company when they applied for compensation. Therefore, as some victims continued to pressure the company with lawsuits, a more militant group of victims and their student supporters started sit-ins at the gate of the Chisso factory on November 1, 1971. This was deeply shocking to conservative local citizens who favored the company. With their support Chisso continued to ignore the claims of the victims who escalated their effort by sit-ins at Chisso's Tokyo office as well.

The victims wanted to arouse public consciousness that money compensation alone was not adequate when their bodies and the environment were being destroyed. Negotiations with the Tokyo office were interrupted when company workers forcibly ejected the victims from the building on Christmas Eve 1971. More violence followed, and in a confrontation at the Goi factory near Tokyo, an American photographer, W. Eugene Smith, was severely beaten because the workers as well as management tried to suppress the negative publicity (206). Victims like Mr. Kawamoto and a supporter were prosecuted and fined for assaulting a worker, but no legal action was taken against employees on the picket line, although one of Mr. Kawamoto's bones was also fractured in the confrontation.

In Tokyo more sympathy was aroused when the Chisso Company placed a steel grating in front of the office to keep out the victims and their supporters. A tent was pitched amid the downtown office buildings, over $100,000 was donated to the cause, and at least 30 people worked each day to maintain this symbol of the victims' struggle over the next year and a half.

While the sit-ins and lawsuits were in progress between 1970 and 1973, Chisso also tried to divide the victims by getting some of them to apply for compensation from the committee of experts assigned to authenticate their claims. This committee generally proposed a lower rate of compensation for new victims who were supposed to have a "light" affliction compared with old victims who were supposed to suffer "heavy" poisoning. The efforts of this committee were thrown into confusion in January 1973, when officials were charged with presenting false affidavits from the victims. Reportedly, this was part of a secret conspiracy by officials and the company to achieve a resolution before the court case was settled.

In March 1973 a judge finally ruled that the Chisso Company was responsible for Minamata disease. He awarded 112 victims and their families $3.5 million and declared that a factory that could not protect the environment should promptly cease operations. The victims were to be paid from 16 to 18 million yen, depending on the seriousness of their symptoms. Then the plaintiffs went to Tokyo to join the group conducting independent negotiations.

The steel grating was finally removed from the Chisso office building, and 500 supporters surrounded it as negotiations continued through the night until the director of the company, Mr. Shimada, promised equal compensation to old and new victims. The final agreement was signed on July 9, 1973, with the Minister of Environment as a witness. Chisso finally agreed to pay an annual stipend and the medical cost of the disease for future as well as current victims. Then the tents were removed.

The government's tendency to suppress information had increased public uncertainty. Thus when news was released that new epidemics of Minamata disease were discovered in the Ariake Sea, Tokuyama Bay, and other places about the same time that the agreement was signed, fish sales plunged as much as 30% in Tokyo. Then thousands of fishermen, merchants, and fish restaurants were reportedly threatened with bankruptcy, whereas red meat sales soared 25% (207). Government efforts to restore confidence only heightened the fear. People did not respond when Prime Minister Kakuei Tanaka ordered his cabinet ministers to embark on a daily fish diet to restore confidence that eating fish was safe. In fact, when the Health and Welfare Ministry published a list of a dozen fish that could be safely eaten, instead of reassuring housewives, sales plummeted. A weekly limit was recommended of 12 small jack mackerel, 2.3 medium-sized cuttlefish, 5.8 mackerel pike, 10.3 medium bass, 6.6 prawns, 1.8 flounder, 1.2 mackerel, and 1.7 scabbard. When this caused a further loss of confidence, the Health Ministry liberalized the list and announced that 46 jack mackerel could be consumed as well as normal portions of fish from the ocean and unpolluted areas. But the price of beef continued to rise, whereas business decreased in restaurants that sold raw and cooked fish.

Professor Ui contends that the public would suffer less social and psychological strain in the long run if all information about pollution problems were made public from the beginning. Instead, the pattern of secrecy is repeated over and over. In the new epidemics medical experts were again called in to deny the symptoms and then to limit recognition to the most serious cases with all the typical symptoms clearly demonstrated. Meanwhile, with 750 confirmed victims, 2000 more remained on the waiting list to be recognized in Kumamoto and Kagoshima prefectures in 1974 as opposing sides argued. The medical experts contended that people falsified their symptoms, and the patients argued that the medical examiners were so highly prejudiced that they could not report the results of physical examinations accurately. Again delaying recognition of claims culminated in litigation by 400 people in Kumamoto District Court.

The list of victims will continue to grow as long as the poor fishermen rely on local fish as a major food staple and the mercury deposits remain in the waterway. Nor are more epidemics being prevented despite a series of laws designed to control air, water, and soil contamination. Their impact is lessened because they conflict with the government's goal of 10% industrial growth per year. In fact, since the pollution victims' claims were recognized in court, a considerable backlash has given further support to industry.

Under a new Pollution Relief Fund Act, polluting industries can contribute money to a fund controlled by the government. Then new victims apply to be compensated from this fund, and the company is saved from the bad publicity of lawsuits. This is really a form of pollution insurance, because the existence of the fund is tacit government acceptance that the company has a right to pollute the environment to some extent. The manager of the Sumitomo Chemical Company was quoted as saying "It is much cheaper to pay compensation after the discovery of the damage from the discharge of pollutants, than to build a complete pollution control system" (208).

Government agencies also are considering direct financial aid to polluting industries. The Chisso Company applied for a low-interest loan to rebuild parts of the factory that were destroyed in an explosion in October 1973. After the economic drain of compensation payments to Minamata disease victims, in 1975 the company needed 5.8 billion yen to rebuild (209). The Federal Department of Commerce debated loaning the company funds but feared that this would set a precedent for other polluting companies that would also expect government aid (209). The Chisso Company subsequently declared bankruptcy. The list of victims still lengthens and is becoming more widely acknowledged, although a Japanese newspaper exaggerated somewhat in saying that "hardly a day passes without some new horror story about bays, rivers, streams, fish and humans poisoned by chemical discharges from Japan's burgeoning industry" (210).

A BOND AMONG VICTIMS

While the ranks of Japanese victims swelled as they battled to force the Chisso Company to accept responsibility, news of the Indians' plight in Ontario, Canada, was brought to their attention. In December 1973 a panel of experts from Sweden, Japan, Canada, the United States, and Finland assembled at Nashville, Tennessee, for a conference on toxic metals in the environment. By then, the mercury contamination controversy had already ended in the United States and Canada. Representatives from the Environmental Protection Agency (EPA) described the new permits soon to be issued under the 1972 Water Quality Amendments (PL 92-500) to regulate dumping wastes into the waterways. The new watchdog agency was also being sued for delays in releasing their first list of toxic contaminants, but enormous sums of money were being poured into research on the cause and cure of various types of contamination.

At the meeting Jun Ui brought the group up to date on new poisonings at Minamata and the efforts of industry and politicians to hide the epidemic or avoid responsibility. In his discussion of Ui's report, Dr. Frank M. D'Itri, in turn, described the mercury-contaminated English–Wabigoon River system in Canada and noted that a new epidemic could be emerging where Indians ate the contaminated fish (211).

D'Itri had tested more fish at Ball Lake in the fall of 1972 while Barney Lamm was transferring his fly-in tourist camp business to Gimli, Manitoba. Lamm wanted to know whether the mercury levels were declining in Ball Lake and what hope there was for the Indians who were again working at the tourist camps. During the visit Dr. Patricia A. D'Itri had examined living conditions on the Indiana reserves, listened to tapes, and read accounts of meetings between the camp owners and politicians. The D'Itris combined their data with Ui's to point out similarities between the Indians' situation at Kenora and the fishermen at Minamata. At that time the Indians had just been given their blood mercury levels from tests conducted in 1970. Frustrated Canadian scientists who had not been allowed to publish their data cheered as red-faced Canadian government and industry officials hastily tried to refute the D'Itris' charges. Ui had not heard about the Indians in Ontario previously, but he immediately began making plans to visit them.

Also in December 1973, after some of Eugene Smith's famous pictures of Minamata disease victims were published in *Life Magazine*, the Lamms had written to Smith and described the mercury problem in Ontario. Then Smith's wife Aileen visited the Dryden–Kenora region in May 1974. One of the Indians asked her what the symptoms of Minamata disease were after he had just described three of them in himself: tingling of the fingertips, constricted visual field, and a heaviness in the back of the head (212). As Aileen Smith returned to Japan determined to help the Canadian Indians, they were also exerting more effort to help themselves.

During the summer of 1974, the Ojibway Warriors' Society occupied Anishnawbeg Park in Kenora. This park named for the Indians is located near a statue of a big silver fish, a tribute to the tourist industry and an appropriate symbol as the Indians charged that the provincial and federal governments were practicing genocide through the willful neglect of the Anishnawbeg, the Ojibway Indians whose food supply had been destroyed by mercury pollution. The Indians were aided by the National Indian Brotherhood, the Society of Friends, and other concerned individuals. Their plight was also finally described by the Canadian Broadcasting Company in a radio documentary entitled "A Clear and Present Danger" (198).

On September 30, 1974, the Ojibway Warriors' Society and Native People's Caravan were also on hand at Ottawa for the opening of Parliament. Despite assurances that they had come in peace to seek justice, both Prime Minister Trudeau and Indian Affairs Minister Judd Buchanan stated that they would not negotiate with violent people, as Buchanan said "Over the barrel of a gun" (198). By then the Indians' violence and illness on the reserves had been tragically documented. A Quaker physician, Dr. Peter Newberry, kept a poignant record of life on the Grassy Narrows reserve when he took up residence there to offer the Indians medical assistance.

Newberry was on hand to greet the traveling team of Japanese visitors when they arrived at Grassy Narrows in March 1975. Ui had gotten financial backing to make the trip from a Japanese newspaper. He pointed out that their visit was front-page news back in Japan, even though it was generally given limited coverage in Canada. The D'Itris and a few crusading reporters joined the Indians and their supporters for a public meeting in Toronto. Besides Ui and the newspaper reporter, the Japanese team included Dr. Masazumi Harada, a physician, Aileen Smith, and Dr. Ken'ichi Miyamoto, an economist.

Dr. Miyamoto remained in Winnipeg when the rest of the group met with the federal Standing Committee on Mercury in the Environment in

Ottawa on March 21, 1975. Therefore, Ui read the economist's statement to the committee. Miyamoto was most struck by the difference in life style and living standards between the white people and the Indians in Canada. He also noted that the extensive pollution of the English–Wabigoon River system had destroyed not only fish but many plants and water animals. Miyamoto cautioned that air pollution was another serious threat to the workers and their families who live just downstream from the main entrance to the plant (213).

At the same meeting Aileen Smith painstakingly translated Dr. Harada's description of the symptoms he had seen among the Indians and compared them with the Minamata disease victims in Japan. Some of the members of the Standing Committee left the meeting before films were shown of human victims in Japan and cats from both countries. As the Japanese cats had shown symptoms of poisoning before the people were stricken, the films showed identical symptoms of poisoning in cats from both countries. One of the Canadian cats had belonged to a man who lived near White Dog Reserve. He fed it entrails when he ate fish. This was the last of several cats that had all begun to act strangely and then disappeared. In some cases they had drown themselves in a nearby creek. When the last cat began to act like the others, movies were taken before the cat was sacrificed. Part of the brain was sent to Frank D'Itri at Michigan State University along with part of another cat that belonged to an Indian woman, Mrs. Keewatin, a widow at Grassy Narrows Reserve. The second cat showed no symptoms of poisoning, but it was sacrificed to determine the background levels of mercury in comparison with the cat that was poisoned.

The cat from White Dog measured 16.9 ppm of mercury in its brain, comparable with poisoned cats at Minamata that ranged between 10 and 20 ppm. Whereas a normal cat's brain would contain about 0.1 ppm, Mrs. Keewatin's cat was hardly representative. Although it showed no symptoms, the cat's brain had 8.6 ppm, also nearly in the toxic range. After brain tissues from both cats were sent to Dr. Takeuchi, he confirmed Minamata disease in the cat from White Dog and the beginnings of degeneration in the brain tissue of Mrs. Keewatin's cat (214).

The Standing Committee members listened courteously to the Japanese, even as Ui expressed shock that the Canadians still did not know how much mercury the Dryden Chemical Company had released into the English–Wabigoon River system. He also wondered why no pressure was being applied to have the firm or the parent company assume responsibility for the damage to the waterway, the victims, or the growing air pollution problem.

The Indians also pleaded their case as this was their first opportunity to appear before the Standing Committee. Tensions were higher when members of the National Indian Brotherhood asked why data on the mercury levels in people, the environment, and the federal cat experiments were not being released. Just before the meeting the group had toured the federal laboratories where experimental cats were developing symptoms of mercury poisoning on a diet of pet food with a much lower concentration of mercury-contaminated fish than cats or people would normally eat. Some of these data were published in scientific journals, but this was no help to the Indians who continued to work at the fishing camps.

Ui persisted in championing the Indians' cause as he had the victims in Japan despite government opposition. He next arranged to have Minamata disease patients invite a group of Indians from the Grassy Narrows and White Dog reserves to come to Japan and see the danger of mercury contamination for themselves (see Figure 14). Teruo Kawamoto welcomed the five Indian men led by Chief Andy Keewatin at a meeting in the Minamata Public Hall on July 22, 1975. As the two groups of victims nervously confronted each other, Ui broke the ice by noting that English was a second language for everyone in the room. The Indians saw the devastation of Minamata disease for themselves, and another Japanese research group returned to the reserves with them in August 1975. They examined Indian guides who were healthy enough to work and noted their symptoms of mercury poisoning.

In the intermediate report of August 17, 1975, the research team again noted the victims' weakness compared with the polluting firms and government. They stressed that the scientists and journalists look for cases of poisoning that show the typical symptoms. Instead, the team was eager to convince the Indians that they should prevent the permanent damage from occurring in the first place if possible.

> Our largest task is to prevent the onset of disease before the discovery of typical victims. Here is clearly the danger of Minamata Disease. There is no doubt of the importance of preventive measures. We must correct our mistake that we limited Minamata Disease only to typical, severe cases for political reasons, and prevent the

LETTER OF INVITATION FROM ALL OF THE MINAMATA DISEASE PATIENTS' ALLIANCE
TO PEOPLES OF GRASSY NARROWS AND WHITEDOG, May 19, 1975.

From last year we have known and heard about you. Until today we
have not been able to do anything to help.

Your bodies have been damaged and even your lives taken from you because
of the poison from the factory waste water. We cannot think of your
problem as a stranger's problem.

We Minamata Disease patients, as victims of Japan's industry and government
and capitalists, have these several decades been discriminated by the
world around us, and have tasted poverty, and physical and mental suffering.

The real dread of mercury can only be understood by those who have had
their bodies made bad, and to their families. And if you rely on the
victimizer (the company), the central government, or the provincial
government, -- they will do nothing for you.

There is no other way to open up the road to help, but for you: Chief
Keewatin, the people of Grassy Narrows, Chief McDonald, and the
people of White Dog, with your own strength, to stand and to speak
your suffering and the anxiety of your sickness to many people.

After eating fish for a long time--and before we realize it--mercury makes
our bodies sick, and our hands and feet to become unable to move, and
not just become numb, paralyzed and have convulsions, but when bad, in
one night can make all of the body no good. It is not just a few who die.

We invite you Chief Keewatin, the people of Grassy Narrows, Chief McDonald,
and the people of Whitedog to Minamata because we want you to see what
the dread of mercury is.

It will be a blessing if our long years of suffering can help you in even
one single way.

And we feel you will understand the dreadfulness of the pouring out of
poisons, and pollution that is spreading over this earth.

If we can help you in changing the anxiety that you have in your hearts
in even one single way--it will be good.

Chief Keewatin, the people of Grassy Narrows, Chief McDonald, and
the people of White Dog, we look forward to the day that we meet you
in Minamata.

                          Representative: Tsuginori Hamamoto

Figure 14. Letter of invitation from Japanese Minamata disease patients to the
Indians at Grassy Narrows and White Dog reserves, May 19, 1975.

damage before it goes to the irrecoverable stage. That is our task here. From this viewpoint, it is not right to limit the problem to the existence of typical victims, even from an administrative or political standpoint, and if journalists limit their activity within this problem, it means that they are leveling down to the lower level of some politicians. The job of journalism should be to see farther than the politician, and give enough warning before hand to prepare a healthy public opinion. (215)

Thus Ui continued to pull no punches as he took scientists and journalists to task as well as officials of government and industry who continue to veil the hazard with secrecy while the poor victims struggle against overwhelming odds. After seeing the hazard for themselves in Japan, the Indians recognized that the Japanese visitors were dedicated to helping the underprivileged, and they publicized Dr. Tsubaki's warning to the guides not to eat the fish (see Figure 15). So the battle in Ontario, like that at Minamata, has escalated insofar as some of the Indians now recognize the danger to themselves and are beginning to speak out against the oppression. Whether they are able to rally enough support to force the government and industry to clean up the pollution and provide them with assistance to seek an alternative life style before they are all destroyed is the next question that must rest on the public conscience, but it is probably already too late for many of them. During the airline strike in the summer of 1976, Jun Ui trudged out of Canada carrying newspaper accounts of confirmed mercury poisoning among Indian children and a growing epidemic on the Canadian reserves in Quebec as well as Ontario.

M E R C U R Y   P O I S O N I N G

Danger to the Guides

THIS IS WHAT DR. TSUBAKI OF NIIGATA UNIVERSITY TOLD US
CONCERNING THE AMOUNT OF FISH AND MERCURY THE GUIDES ARE
EATING WHEN WE TALKED TO HIM AT NIIGATA UNIVERSITY.

(DR. T. TSUBAKI IS A PRESTIGIOUS NEUROLOGIST IN JAPAN.
HE IS ALSO AN ADVISOR TO WHO, THE WORLD HEALTH ORGANIZATION
OF THE UNITED NATIONS, AND IS KNOWN INTERNATIONALLY AS AN
EXPERT OF MINAMATA DISEASE.)

ACCORDING TO DR. TSUBAKI:

EVERY TIME A GUIDE EATS FOUR FILLETS (ONE POUND OF FISH)
WITH A MERCURY LEVEL OF 2 ppm HE IS TAKING IN 900ng OF MERCURY.
IF HE EATS THIS AMOUNT OF FISH EVERY DAY, IN ONE WEEK HE WOULD
HAVE EATEN 6.3 mg OF MERCURY.

SO: THERE HAVE BEEN PEOPLE WHO HAVE STARTED TO SHOW SYMPTOMS
     WITH AS LITTLE AS 20 mg OF MERCURY IN THEIR BODY. (THAT
     HAPPENED IN IRAQ.)

AS YOU CAN SEE, IF A GUIDE EATS 6.3 mg OF MERCURY A WEEK,
IN FOUR WEEKS HE WOULD HAVE EATEN 25.2 mg OF MERCURY. THAT IS,
IN ONE MONTH HE HAS EATEN MUCH MORE THAN THE 20 mg THAT HAVE
GIVEN SOME PEOPLE SYMPTOMS.
MERCURY DOES GO OUT OF THE BODY, BUT AT A MUCH MUCH SLOWER
RATE THAN THE AMOUNT OF MERCURY THAT THE GUIDES ARE EATING.

SO EVERY DAY AND EVERY WEEK, MORE MERCURY GETS PILED UP
IN THEIR BODY.

EVERY DAY COUNTS. THE SOONER THE GUIDES STOP EATING THIS
MERCURY CONTAINING FISH, THE SOONER AN OUTBREAK CAN BE
STOPPED.

IT IS A KNOWN FACT IN MINAMATA AND NIIGATA THAT OFTEN IT TOOK SOME
YEARS OF MERCURY PILING UP IN THE BODY TO FINALLY CAUSE BAD SYMPTOMS.
WE KNOW THAT THIS MERCURY PROBLEM WAS MADE PUBLIC IN 1970. THIS YEAR
IS THE FIFTH YEAR. THIS COULD WELL BE THE YEAR THAT THE OUTBREAK
COULD START. WE MUST WORK QUICKLY TO STOP AN OUTBREAK!!

ALSO: REMEMBER THAT EVEN IF THERE IS POISON IN THE FISH, THE FISH
       DOES NOT TASTE BAD. THAT IS WHAT'S SO SCARY ABOUT MERCURY
       POISONING.

Figure 15. After the Indian delegation talked with Dr. T. Tsubaki of Niigata University in August 1975, they issued this warning to Indian guides not to eat fish taken from the Wabigoon-English River system.

101

II

TECHNICAL
APPLICATIONS:
ANCIENT AND MODERN

8

HISTORICAL BACKGROUND

The inorganic mercury compounds include the liquid element and its vapors as well as the mercurous and mercuric ions and their salts. Without heat nitric acid dissolves mercury to form mercurous salts, whereas adding heat and excess nitric acid forms mercuric salts. Elemental mercury gradually vaporizes at ordinary temperatures, although it is relatively stable in the presence of air, oxygen, carbon dioxide, nitrous oxide, and ammonia. Mercuric oxide forms when the element is heated at moderate temperatures, and the oxide decomposes into mercury and oxygen at temperatures greater than 500°C. In 1777 Antoine Lavoisier used this reaction to prove that oxygen is a component of the atmosphere and to refute the theory that phlogiston evolves when substances are burned.

New technology has altered the patterns of exposure to mercury poisoning over time. Among ancient procedures that continue with modern methods, mercury is still mined, refined, and amalgamated with other metals such as gold, aluminum, and zinc to separate and purify them (216). However, other applications have decreased or become obsolete. Some early statistics were compiled in England where the first protective legislation concerning mercury was passed in 1878.

After the British Parliament passed the Factory and Workshop Act, young persons were prohibited from "silvering" mirrors with mercury. And after 1897 physicians were required to report industrial illnesses caused by mercury. Thus in the same year, a factory inspector reported that a woman had to stop working in a scientific instruments factory because of ill health caused by exposure to cold metallic mercury. A stalemate ensued when factory administrators countered that she had actually been fired for making injurious statements about them. However, the incident prompted the Home Office Dangerous Trades Committee to examine other hazardous industrial processes to determine if the workers needed further protection. Then the Inspector of Factories required reports of deaths from industrial mercury poisoning. Health authorities subsequently listed workers who produced scientific instruments, a few who still silvered mirrors, and many casualties among furriers. Until mercuric nitrate was replaced in the carrotting fluid to felt fur for making hats, this long-standing use of mercury caused thousands of casualties.

When mercury poisoning became a legally notifiable disease under the Factories Act of May 1, 1899, this was not considered an innovative piece of legislation in Great Britain because of precedents in other countries.

Nearly all civilized countries have legislated to a greater or less extent for workers in white lead, mercury and phosphorus. By this action the various Governments have recognized that ill-health follows unregulated employment in certain trades, while the compulsory medical examination which many countries in addition require is an expression of opinion that by timely detection much of the evil consequences of industrial poisoning can be prevented (217).

This system of legal notification enabled authorities to chart the changing patterns of mercurialism. For instance, in 1912 the Chief Inspector of Factories recorded seven cases of mercurialism among gilders during the first 6 years that notification was required. This trade has since decreased as few members of royalty are now building elaborate palaces. Only five more occurred by the end of World War II and two after that, the latest in 1961 when a gilder was overcome by mercury fumes in London after some emerging African nations submitted rush orders for gilded articles of state authority such as a large golden mace (218–219). Dusting fingerprints with a mercury-containing powder was also a very specialized source of contamination until this process was halted. Other applications are more sporadic. During World War II over 334,000 lb of fulminate of mercury were manufactured for detonators and blasting caps, whereas less than 32,000 lb are required annually in peacetime (220).

As old applications of mercury declined, new ones with larger requirements were developed. In 1923 Mercurio Europeo, the mercury cartel, offered a 5000-pound sterling reward for any new mercury use that would require 5000 flasks (76 lb each) annually by 1935; no one claimed the reward. Nonetheless, by 1933 eighty industries required mercury, and the list had grown to over 3000 by the 1970s (221). In the United States 884 companies used mercury by 1969 with the bulk consumed by 80 companies, most of them located in the eastern part of the country (222). As incidents of mercurialism from felting decreased after the turn of the century, workers were more often exposed while manufacturing and repairing mechanical and electrical equipment. These industries have grown steadily along with requirements for mercury as an intermediary in the manufacture of other products, particularly plastics, chlorine, and caustic soda. The relative changes in the mercury-consumption pattern for individual American industries during the 1950s, 1960s, and 1970s are presented in Figure 16.

Despite the fact that the Chief Inspector of Factories and the British Medical Association recommended routine monthly physicals in 1899, only the most severe cases of mercury poisoning were reported, between 500 and 1000 by 1902. Thus in 1905 when 40% of the employees in a thermometer factory had mercury burns on their hands, this was not considered serious enough to notify authorities.

The early symptoms of inorganic mercury poisoning are also so general that they are difficult to detect. A worker may suffer headaches, fatigue, nervous anxiety, restlessness, and a loss of appetite, or he may simply become unusually timid, embarrassed, depressed, and discouraged. As his irritability and restlessness increase, he may lose his ability to concentrate. As his memory declines, his fear and indecision intensify (223). However, all these symptoms are reversible. Although the inorganic mercury compounds concentrate in the kidney, liver, and spleen, they are readily excreted and do no damage unless the organ's threshold tolerance level is exceeded.

Unlike methylmercury which concentrates during a period of latency in which no symptoms are observed and then may strike a crippling or even lethal blow to the mind or body, the symptoms of inorganic mercury poisoning continue to develop gradually and may be reversed if the exposure is halted soon enough. The first clear physical symptoms are numbness of the fingers and toes and then of the lips and tongue. Among industrial workers the most common symptoms have been a sore mouth,

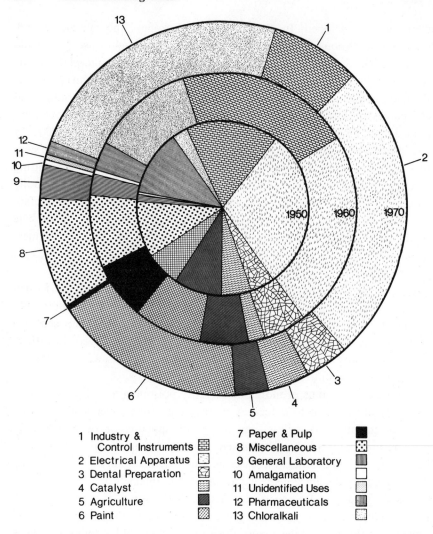

1 Industry &
      Control Instruments
2 Electrical Apparatus
3 Dental Preparation
4 Catalyst
5 Agriculture
6 Paint
7 Paper & Pulp
8 Miscellaneous
9 General Laboratory
10 Amalgamation
11 Unidentified Uses
12 Pharmaceuticals
13 Chloralkali

Figure 16. The relative changes in mercury-consumption patterns for individual American industries during the 1950s, 1960s, and 1970s.

muscular tremors, and psychic irritability. All or only one or two may signal excess mercury in the human body. The intention tremors first appear only when the victim exerts himself in some unusual way or has attention drawn to his condition. Although the worker is steadily more incapacitated in other surroundings, he can often continue to perform familiar routines normally. Consequently, workers were often reluctant to call attention to their illness for fear that they would lose their pay or be fired for making a complaint. They became more willing to speak up after mercury poisoning was included in the Workman's Compensation Act of 1906. Casualties also declined where medical authorities identified workers suffering from mercurialism and then checked their coworkers to prevent more exposure. Thus protective legislation and preventive medicine decreased the number of instances in which the symptoms of inorganic mercury poisoning intensified until death was caused, often from uremia because the kidneys were damaged until passing urine became impossible (224–227).

If death is not caused by uremia and further degeneration occurs, the tremors can intensify into spastic motions and then rigidity. Emotional deterioration begins with a loss of memory and may degenerate into melancholia and hallucination. When the individual loses his speech and hearing, it is difficult to remain in contact with people around him even if his mind is not impaired. Thus some patients have spent years in mental hospitals, although they were fully aware of their situation. Careful, time-consuming efforts to establish contact confirmed their possession of their mental faculties. Essentially the same symptoms develop more gradually when the exposure is chronic rather than acute. Thus it is possible for long-term exposure to produce the gradual mental and physical deterioration that is usually associated with senility.

After 1936 British health authorities concentrated their efforts on preventing rather than detecting mercury poisoning. To determine if workers were concentrating the element, they were checked for water retention, albuminuria, proteinuria, and mercury in the urine. But monitoring techniques continued to be inadequate, and routine medical examinations were more apt to detect than prevent poisoning even after the 1963 Mercury Processes Regulation was passed (218). In 1974 agitators even charged that federal hygiene standards should be abolished, because health inspectors placed conformity to regulations in higher priority than protecting workers' health in Great Britain.

Despite changing requirements and a shift in emphasis from detecting to preventing poisoning in industry, the early protective legislation in England offers much more information on early conditions than is available in the United States. Federal hygiene standards were not imposed on industries using mercury until 1940, and notification of mercury poisoning is still not required. However, maximum allowable concentrations of mercury have been set and are being changed as more information about the effects of exposure become available. In 1959 the First International Symposium set the first international standard for maximum allowable concentrations (MAC) of mercury vapors in occupational exposures. The acceptable level was then defined as:

> . . . that average concentration in the air which causes no signs or symptoms of illness or physical impairment in all but hypersensitive workers during their working day on a continuing basis as judged by the most sensitive internationally accepted tests (223).

Ten years later, the mercurials were divided into three categories with MAC values subject to change when more data became available (223). Around the same time, higher standards were also adopted in the United States. A maximum threshold limit value (TLV) of 0.1 mg of mercury vapor and inorganic mercury compounds per cubic meter of air had been the standard set by federal authorities in 1941 to prevent workmen from concentrating more than 0.25 ppm of mercury in their urine. In 1967 the American Conference of Governmental Industrial Hygienists acknowledged the effects of different types of mercurials on the human body by recommending a maximum of 0.01 mg of organic mercury compounds per cubic meter of air (228–229) and 10 μg of inorganic mercury per cubic meter in 24 hours with a 50-μg limit for an 8-hour day.

The new standards still did not take into account the possible damage from chronic exposure. Thus the National Institute for Occupational Safety and Health (NIOSH) subsequently recommended a maximum of 0.05 mg/m^3 concentration of mercury vapors for an 8-hour day and 40-hour week with skin contact limited to 10 μg/m^3. The maximum acceptable urine excretion was reduced from 0.25 to 0.15 mg/1 with an adjusted specific gravity of 0.024 (230–232). This level was considered safe because people with no daily exposure normally might have this much mercury in their urine.

As the acceptable levels of mercury have been lowered and separate standards have been set for the more hazardous organic mercury compounds, workers in mercury-related industries can now be carefully monitored because of more sophisticated equipment to confirm the levels of exposure in the environment and the concentration of mercury in their bodies. Therefore, inorganic mercury poisoning is a preventable illness if the proper sanitary conditions and health precautions are maintained.

Table 1 summarizes the industries, present and past, that may expose workers to mercury poisoning and discharge the element into the environment (233).

Table 1. Industries with a potential Mercury Hazard, Present and Past (Courtesy of P. A. Neal and the American Journal of Public Health)

Industry		Description	Principal Method of Entrance Into the Body
Mining	Cinnabar (HgS)	Hazard depends upon the process used; hazardous occupations are those connected with condensers, purification, flask filling, distillation repairing and cleaning	Inhalation of vapors and dusts
	Gold	Amalgamation with mercury, closed process used at present	Inhalation of vapors
	Silver	Amalgamation with mercury	Inhalation of vapors
Chemical manufacture	Manufacture of organic and inorganic mercury compounds	Inorganic mercurous and mercuric compounds, as chlorides, nitrates, iodides, cyanides; and organic compounds such as mercurochrome, metaphen, dimethyl and diethyl mercury compounds	Skin, ingestion, and inhalation of vapors and dusts
	Mercury as a catalytic agent	Starting with acetylene, using salts of mercury as catalytic agent, to produce synthetic alcohols, ketones, and acids; volatile toxic organic mercury compounds also produced Sulphonation, oxidation (Kjeldahl) bromination, nitration, alkylation (Friedel-Crafts)	Inhalation of vapors

111

Table 1. (cont'd)

Industry	Description	Principal Method of Entrance Into the Body
Cyanogen	*Cyanide of mercury decomposes by heat liberating cyanogen and mercury; mercuric oxide plus hydrocyanic acid gives mercuric cyanide*	*Skin contact; Inhalation of vapors*
Chlorine	*Electrolytic process--Solvay process with sodium amalgam*	*Inhalation of vapors*
Alloy Makers, including amalgam makers	*Mercury forms amalgams with most metals readily. It does not form an amalgam with iron, nickel, aluminum, cobalt and platinum*	*Inhalation of vapors*
Electric		
Standard cells	*Mercurous chloride and metallic mercury*	*Inhalation of vapors*
Mercury switches		
Mercury vapor lamps		
Electric contacts and insulators		
Electrodes	*Amalgamation of zinc*	
Accumulators	*Amalgamation of zinc*	
Incandescent electric lamps	*Production of the vacuum, rarely used at present*	
Electric meters		
Electric furnaces	*Sealing induction furnaces with mercury*	
Radio tubes	*Production of vacuum by mercury*	

Table 1. (cont'd)

Industry		Description	Principal Method of Entrance Into the Body
Manufacture of felt hats	Fur cutting	Carrotting solution composed of a solution of mercury nitrate. In the U. S. practically all employees exposed to a measurable concentration of mercury vapors	Inhalation of vapors and dusts
	Felt hat industry	From the mercury nitrate solution used in carrotting the fur, chief occupations with mercury exposure are: mixers, blowers, coners, devil operators, hardeners, sizers, starters, stiffeners	Inhalation of vapors and dusts
Dentistry		Treatment by heat of mercury amalgams especially silver and copper amalgams	Inhalation of vapors
Laboratories		Manipulation of mercury Spilled mercury accumulates on floors, work benches, etc.	Inhalation of vapors Skin
Pharmaceutical		Manufacture of antiseptics, cathartics, diuretics, disinfectants, bactericidal agents, ointments and solutions containing mercury or mercury compounds, both organic and inorganic. Spilled mercury	Inhalation of vapors and dusts

Table 1. (cont'd)

Industry		Description	Principal Method of Entrance Into the Body
Manufacture of	Vermicides Insecticides Fungicides	Organic and inorganic compounds of mercury. Mercuric chloride used in wood impregnation	Skin Inhalation of dusts
Explosives	Detonators Detonating fuses Percussion caps Pyrotechnics	Mercury fulminate in copper or brass tubes mercuric fulminate used mercuric thiocyanate used in Pharaoh Serpents Mercuric sulfate used as chemical catalyst	(Dermatitis) and Inhalation of dusts and vapors
Decoration of chinaware and porcelain		Mercuric oxide and mercurous chromate used for decorating chinaware and porcelain	Inhalation of vapors Irritation of skin
Photography		Mercuric chloride and mercuric bromide used as intensifier Mercuric double rhodanide	Inhalation of vapors Absorption from skin
Printing Trades		Pantone process employs a paste with mercury as a basis for cleaning silver typographic plates as well as for typographical operations. Vermilion is used in some inks. Production of printing plates. Color printing as used in linoleum	Inhalation of vapors

Table 1. (cont'd)

Industry	Description	Principal Method of Entrance Into the Body
Battery	Dry battery manufacture; heat used in melting and soldering, volatilization of the mercury. Mercuric sulfate used as a chemical catalyst in electric batteries. Lead mercury solder	Inhalation of vapors
Paint	Mercury is a constituent of anti-fouling paints for ship bottoms. Tin-bismuth-mercury amalgam as a paint mixed with oil. Preservative in water based paints	Inhalation of dusts and vapors
Engraving, embossing	Mercuric chloride used in etching or steel. Damascening of guns and swords mercuric nitrate used	Inhalation of vapors and dusts, skin
Mercury vapor boilers	A process of power generation which uses mercury instead of water in a boiler. Use of mercury vapor for heat transfer in operations such as chemical reactions or distillations, as used in the petroleum industry	Inhalation of vapors

Table 1. (cont'd)

Industry	Description	Principal Method of Entrance Into the Body
Jewelry	Production of imitation jewelry, copper gilded by the mercury process, or by plating process. Waste, debris, etc., washed and triturated in a mill with mercury, forms amalgam, then freed from mercury by heat. Metallic mercury. Heated mercury amalgams	Inhalation of vapors
Shooting galleries	Mercury vapors liberated from use of mercury fulminate percussion caps	Inhalation of vapors
Taxidermists	Mercuric chloride used to treat skins for conservation	Skin and Inhalation of vapors and dusts
Anatomical preparation	Tin, lead, bismuth amalgam injection	Dermatitis
Embalmers	Mercuric chloride	
Backing of mirror	Mercury very little used at present	
Glass globe coating	Mercury used in coating	
Bronzing	Mercury may occur in bronze as a contamination of tin	
Cosmetics	Mercury compounds used in the preparation of hair lotions and freckle preparations	

116

Table 1. (cont'd)

Industry	Description	Principal Method of Entrance Into the Body
Hardening process	Hardening tools in mercury	
Dyeing of woolen goods	Mercuric nitrate in dyeing woolen goods	
Manufacture of aniline red	Mercuric nitrate in making aniline red	
Printing of cotton goods	Mercuric chloride used as a mordant	
Artificial flower making	Mercuric iodide used for coloring	
Color manufacture	Vermilion used as an inorganic dye	
Mordants for seed	Chlorphenolmercury nitro-phenolmercury, methylmercury	
Fuchsin	Mercuric chloride used in production	

9

MINING AND SMELTING:
THE BASIC INDUSTRIES

THE RICHEST MINES

Whereas all soil normally has a minimum of 50 ppb of mercury, an irregular belt of much richer ore deposits encircles the earth (234) (see Figure 17). Although mercury is less abundant than platinum, uranium, and silver, a few readily accessible ore deposits make the liquid element less expensive to obtain. Almadén, Spain; Monte Amiata, Italy; and Idria, Yugoslavia, are the three richest mercury mines in the western world. Almadén, a name that means "the mine" in the Spanish Sierra Morena mountains is so rich that liquid mercury sometimes trickles out of the walls. In the San Pedro vein the cinnabar ore has averaged 20% mercury (235).

The deposits of mercury formed relatively recently in the earth's history when hydrothermal solutions from hot springs or volcanic activity penetrated unstable geological formations to replace porous sandstone or limestone formations with mineral solutions containing mercury. They became veins, impregnations, stockworks, or replacement lodes of mercury ore in quartzite or other minerals. Although cinnabar and metacinnabar ore are usually richest in mercury, the element combines with at least 30 minerals to form compounds such as selenides, sulfides, and oxides (234). Mercury

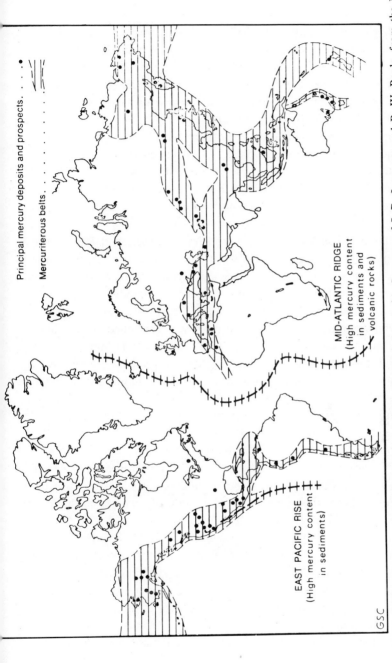

Principal mercury deposits and prospects. . . . . •

Mercuriferous belts. . . . .

MID-ATLANTIC RIDGE
(High mercury content
in sediments and
volcanic rocks)

EAST PACIFIC RISE
(High mercury content
in sediments)

GSC

Figure 17. Generalized map showing the mercuriferous belts of the world. Courtesy of I. R. Jonasson and R. W. Boyle, from the Proceedings of the Symposium on *Mercury in Man's Environment*, Royal Society of Canada, February 15–16, 1971, Ottawa. Canada.

is often combined with pyrite, macasite, chalcedony, quartz, calcite, dolomite, ankerite, and stignite (234–235).

No one knows how long ago a volcanic eruption poured out the thermal solution that replaced quartzite with cinnabar ore at Almadén. However, Phoenicians were carrying out cinnabar ore as early as 700 B.C., and the ancient Etruscans mined at Monte Amiata. These two most ancient mines have both been worked since prehistoric times. As the Romans extended their empire, Monte Amiata in the northern part of the Italian peninsula came under their control by 295 B.C., and Almadén was captured from the Carthagenians at the end of the Second Punic War (218–201 B.C.). Then Roman authorities closed the Italian mines and imported mercury from Spain, perhaps to protect Italian workers from mercury poisoning or to prevent the devastation of the Italian landscape.

The Italian mines were not reopened for centuries, and Roman slaves and prisoners were sentenced to labor at Almadén. Thus mercurialism was a common disease among slaves. Since the miners' life expectancy was only 3 years, Justinian described a sentence to work at Almadén as almost equal to a death sentence, and Plutarch chastized a mine operator for employing slaves who were not criminals. Slave labor was continued when Almadén was returned to Spanish rule in the twelfth century A.D. until the middle of the eighteenth century. Thus in 1719 Antoine Jussieu observed the poisoning among slaves who worked and ate in the mines and noted that local residents were not affected by the fumes. He marveled that nearby crops and trees grew normally and that springs near the mines yielded potable water.

The Spanish sold mercury from the rich mine to finance military and colonial expansion. King Alfonso VIII awarded a part interest in Almadén to the Order of Calatrava, a group of militant monks who earned the King's gratitude for their part in the reconquest of Spain from the Moors. Although the original award was proffered in 1168 A.D., the Cistercian monks did not actually take possession until 1249 when Ferdinand III (1199–1252 A.D.) renewed the offer. The monks operated the mines until 1348 A.D. when Alfonso XI (1311–1350 A.D.) again asserted royal domination. The government then alternately operated the mines and leased them to private enterprise for a percentage of the profits.

Spain was the world's greatest exporter of mercury for centuries. By 1795 Almadén had been exporting quicksilver for at least 2287 years and had produced in excess of 500 million lb despite occasional fires and the constant threat of mercury poisoning. A fire in 1550 A.D. caused the suspension of mining operations for nearly 2 years, and production did not return to normal until 1555. Another fire broke out in 1755 A.D. and burned for 30 months. Production was impaired until 1760.

After paid labor was introduced, the work schedule was revised several times to decrease mercurialism among the miners. First the hours were limited to six a day, 3 days a week (236). Then a schedule of 8 work days a month with 4½-hour days was instituted, but mercurialism continued to decimate the work force. The rich veins were a curse to the workers since the liquid element exhausted more vapors than the cinnabar ore, and antiquated metallurgical processes were an even greater hazard. In 1921 twenty-one hundred miners and 400 furnace men were stricken, primarily by mercury vapors escaping from the furnaces (237). Then the work schedule was again altered to 6-weeks exposure followed by an equal rest period. Unless this was enforced, symptoms such as trembling and vertigo usually appeared within 4 months. Similar horrors were reported in Italian mines after they were reopened.

In 1920 at Monte Amiata an Italian miner was so impregnated with mercury after working 6 months that he could turn a piece of brass white if he held it in his mouth or handled it with his fingers (233). But the most extreme case was a Spanish miner who suffered slightly when he mined cinnabar but developed continuous tremors when he worked in a reduction plant, because vapor levels were even higher where the ore was converted into mercury. Sometimes this man found it impossible to feed himself. At one point he suddenly began to have violent spasms of the right leg. It "began to work up and down like a pile driver." The man's thigh would draw up to his abdomen with the leg flexed. Then it would kick out again so violently that his fellow workmen could not hold it still. The leg jerked back and forth rhythmically about 60 times per minute, and drugs had no effect for the first 3 days of the week that this extreme reaction lasted. By the time the spasm ended, the miner was exhausted. He had only been able to consume liquids, a little at a time, because he could not remain still long enough to swallow more. Although he then stopped working with mercury, he remained a prey to headaches, insomnia, bad dreams, depression, and irritability. His tremors returned when the man was excited, but never again with such severity (237).

IDRIA

The third major European mercury deposit was discovered at Idria in what is now Yugoslavia around the time Columbus returned from America, just before mercury came into great demand to treat the first great syphilis epidemic. A lifesize picture of the Roman god Mercury in his winged helmet now stands at the head of the stairs leading to the small museum that commemorates the town's mercury mining history. The collection of outdated mining equipment shows the mechanical improvements since mining began in the fifteenth century.

The rich mercury deposit was discovered by accident after a cooper left a wooden bowl in a stream to moisten. Many days later when he returned, the bowl was too heavy to lift because the bottom was filled with a silvery liquid. Although he did not know what the substance was, the cooper carried his bowl to a neighboring town in hopes that it might be of value. Then a soldier accompanied him back to the site of the third richest cinnabar deposit in the world. The enterprising barrel maker subsequently founded a mining company that was soon confiscated by the local lord. After 1580 the mine was claimed by the prince of the province, and its ownership was contested in subsequent wars. Although part of the Austro-Hungarian state until 1918, Idria was then ceded to Italy and became part of Yugoslavia after the Second World War (238).

At first only elemental mercury was extracted from the rich bituminous shale, but soon cinnabar deposits were also mined and smelted by a very primitive process. Alternate layers of ore and wood were covered with earth and set afire. Then the liquid mercury was reclaimed as it flowed onto the ground. The first mass poisoning was reported at Idria in 1553. Paracelsus studied mercurialism among the miners and incorporated his observations in a book on their diseases. Even then, proposals were not lacking for ways to protect their health. A law requiring a 6-hour work day was instituted in 1665, reportedly the first legislative measure passed for industrial hygiene in history. But the miners' tremors continued, and Giovanni Scopoli described them again in 1761. To prevent mercurialism, a local doctor who practiced between 1766 and 1779 recommended taking hot baths and drinking milk.

Exposure increased in 1785 when Austria concluded a trade agreement with Spain that called for greater mercury production. Then the mines were worked year round instead of only during the 3 winter months.

Although the mining equipment was gradually mechanized and vertical flame furnaces were constructed in 1789, all 900 inhabitants of Idria again were exposed to mercurialism after a fire broke out in a mine tunnel in 1893 (239). Townspeople were less severely afflicted than the miners, but even domestic animals suffered. The cows salivated and became weak. Calves often were aborted or died soon after birth (237).

In 1897 the workers were scheduled to tend the furnaces for 1 month and then rotate to less hazardous tasks for 2 months. Consequently, mercurialism declined from 122 cases in 1898 to five in 1908. And a variety of major precautions have helped improve safety since then. Until showers and places to change clothing were provided, the men were also exposed to mercury vaporized from soiled clothing or absorbed through the skin when they perspired. Children were said to experience mercury poisoning if they slept in the same bed with fathers who worked in the mines. Cleanliness and special work clothes helped prevent this. As the miners tunneled further underground, the air was more carefully monitored and the mine shafts were ventilated. Moreover, in 1950 wet drilling techniques eliminated the danger of dust from blasting.

A major improvement was to move the roasting furnaces from the floor of the valley to a refinery on top of a nearby hill. Now the ore is transported from the mine by buckets on a revolving mechanical hoist, and vapors from the furnaces are discharged away from the town. The miners' urinary mercury levels are also carefully monitored, and since the men can be moved to jobs without exposure when necessary, they now work 8-hour days, 6 days a week. The land stays green, and townspeople seem to suffer no ill effects from the little mercury that translocates into growing vegetables, although drops may even ooze out of the ground occasionally to collect on the grass or foliage (240).

THE NEW WORLD

In the Western Hemisphere the major mercury deposits are located on the East Pacific Rise (see Figure 18). Ancient Indian tribes in South America labored in the Peruvian mines, whereas North American Indians came from as far away as the Columbia River Basin to obtain cinnabar from the mountains around Santa Clara, California. They ground the red ore into vermilion powder and mixed it with water to make their war paint.

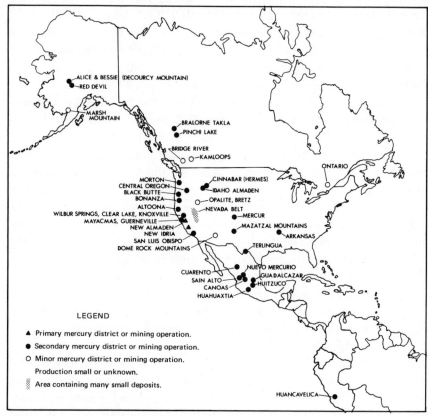

Figure 18. The principal mercury deposits in North America.

When the Spanish arrived in Peru, they recognized the red substance with which the Indians decorated their faces and began to mine the cinnabar ore. Demand for the element escalated when silver was also discovered in the New World.

On March 4, 1552, the governing princess of Valladolid, Spain, acknowledged an urgent request from Mexico for mercury to process silver. Subsequently, the element was transported from Almadén to Mexico, Peru, and Bolivia to recover gold and silver; and some was also exported from Peru to Mexico. Later, when Peruvian mercury deposits on Mount Huancavelica were wedded with Bolivian silver deposits on Mount Potosi, the "marriage" of the two was duly celebrated. The Bolivian Viceroy subsequently offered Pedro Fernandes de Velasco a reward for bringing the patio process from Mexico to Bolivia (241).

Bartolome de Medina had introduced the patio or cold amalgamation process of refining silver at the Mexican silver mines in Pachuca about the middle of the sixteenth century. The pulverized silver or gold ore was mixed with salt brine on a paved floor or patio. Then a mixture of roasted copper, iron pyrites, and elemental mercury was added, and workmen with hoes and rakes blended all the components. Finally, the liquid silver or gold amalgam was removed and heated to recover the silver or gold and mercury. A more efficient hot amalgamation process was being employed by 1640 (242).

Despite the proximity of silver refineries in Bolivia to Peruvian mercury mines, much of the liquid element continued to be imported from Spain. Navigating a long sea route was easier than transporting mercury up and down mountains by mule in the South American countries. Consequently, ships transported mercury across the ocean under a Spanish crown monopoly that endured 300 years from 1560 to 1860. Some cargoes were lost as vessels floundered along the coastlines of the Americas (243), but most of the mercury extracted silver that "hath come out of this Kingdom (and) hath filled the world with riches and admiration."

Because mercury was needed to refine Mexican silver, Spanish authorities sought new deposits that could be mined and transported more conveniently. Thus when a cinnabar ore discovery was reported at Lodos Prietos on Laguna Hill in the California territory in 1796, the Mexican governor ordered the commandante of the military post at Santa Barbara to send a small detachment of men under the command of Sargent Jose Maria Ortega to check on the story. They confirmed the discovery, but no mercury was mined, and the deposit was not mentioned in accounts of a mill built to refine silver on a nearby stream in 1824. In 1845 Captain Andres Castillero took out a mining claim for either gold or silver, although he also sought the "Cave of the Red Earth" where Indians occasionally still pilgrimaged to obtain red and yellow pigment. These Indians lived near the old Santa Clara Mission about 50 miles from San Francisco, California. In the Santa Clara Mountains Captain Castillero rediscovered the richest vein of cinnabar in what later became the United States. By heating the ore in his gun barrel until the liquid globules of mercury ran out, Captain Castillero also became the first known mercury poisoning victim at New Almaden (244).

Castillero confirmed his rich discovery by sending samples of ore back to the College of Mines in Mexico City where he had previously studied chemistry and metallurgy. The samples averaged 35.5% mercury, and some specimens were practically pure cinnabar. The Captain then declared his intention to mine mercury and claimed the $100,000 reward offered by the Mexican government to anyone who discovered a mercury mine as rich as Almadén. A limited mining operation was begun with a few Indians under the supervision of a mayordomo and a blacksmith. After the small furnace exploded, very little mercury was produced by heating the ore in kettles or gun barrels.

Despite this early Spanish claim, in the 1850s James Alexander Forbes was given title to the mine in a lawsuit instituted after California became part of the United States. Forbes had begun a larger mining operation in 1846. Two years later, the Reverend C. S. Lyman noted the extensive mining operation that had been mounted by Barron, Forbes, and Company at New Almaden despite an initial delay when the vessel full of equipment bound for the Mexican territory was confiscated by United States forces. Forbes had then recruited miners and purchased tools in Mexico (245). Reverend Lyman's excitement over the mercury mine prevailed over a casual note that gold had also been found recently on the Sacramento River near Sutter's Fort where "it occurs in small masses in the sands of a new mill race and is said to promise well" (246).

As the Gold Rush escalated in 1849, mercury was mined nearby to separate the prospectors' newly panned gold from its ore. The mercury and gold ore were placed in riffled tables and washed in a stream of water. Some mercury amalgamated with the gold, and more was lost into the water with the tailings. The prospectors inhaled the mercury fumes as they roasted the amalgam in their frying pans until only the gold remained. Thus mercurialism was as common to gold prospectors as it was to mercury miners.

Sam Williston, head of the Quicksilver Institute, estimated that up to a million flasks of mercury may have been utilized to obtain gold in the Sierra Nevada foothills. Consequently, millions of pounds were discharged into streams and rivers that drained into the San Joaquin and Sacramento Rivers (247). This may elevate the mercury levels in catfish in the Merced River and in seals off the coast of Santa Rosa, California (248). If the mercury wastes also spread along the coastline, some evidence indicates that it may also have contaminated the fish, seals, and walruses on which the Eskimos depend for their livelihood.

The early California mercury miners carried back-breaking 200-lb sacks of ore up narrow, hazardous steps that extended farther as tunnels were dug into the sides of the mountains. But vapors exhausted while smelting the ores were a more immediate health hazard among workmen compelled to remain in close proximity to the furnaces and condensers. Persons with delicate nervous systems were said to be particularly subject to the injurious fumes. Consequently, ladies reportedly salivated when casually passing by, and miners' wives were poisoned by washing their husbands' mercury-permeated overalls. Vegetation was usually not affected beyond the immediate vicinity of the chimney, but if cattle were allowed to browse in the area during the dry season, they would sometimes salivate and die (244).

Miners' folklore in Spain and California suggested ways to combat mercury poisoning. In both places alcohol was believed to increase the physical and emotional symptoms of poisoning. Old hands advised new men to give up smoking and to chew tobacco instead, because mercury vapors could be inhaled with cinnabar dust on the tobacco in a cigarette. Some miners believed extensive spitting could remove the inhaled vapor. And "sweating the poison out" has continued to be recommended (237). Thus some mines still provide steam rooms in which workmen try to rid themselves of mercury by perspiring excessively.

Exposure from furnace exhausts could be decreased by extending the flue and building a tall chimney to condense the mercury and arsenic fumes (249). By 1854 larger brick furnaces had been built to extract mercury from cinnabar ore without releasing vapors except through the condensers. Nonetheless, enough arsenic and mercury fumes still escaped until workers could only spend one week out of four stoking the ovens, cleaning the condensers, and rebricking the tops of the furnaces (244). By 1864 a furnace had been erected that could reduce from 150,000 to 200,000 lb of ore at a single charge with from 4 to 6 charges per month. Despite being built on "the most approved principles," the odor was strong, and workmen suffered the usual physical symptoms.

Among the diverse emotional reactions were radical alterations in the individual's personality. At one extreme, a young man who was a member of a small group of mercury producers in California became so apathetic and drowsy that he fell asleep at a meeting where the most crucial questions about the fate of the company were being decided (237). Other victims often became shy and excitable at first and then very irritable. Some of them had to give up work because they could no longer take orders. Foremen would become too impatient with their men to remain in charge.

AMERICAN MINES SINCE THE TURN OF THE CENTURY

Between 1875 and 1882 the mine at New Almaden in Santa Clara County averaged 18,000 flasks of mercury per year, nearly half the total output from California (250). This state produced and consumed most of the mercury in the United States until the beginning of the twentieth century. Since then, California has produced up to 30,000 flasks per year with slightly over 50% from underground mines.

Because many marginal mining operations have only opened when the price of mercury was high, they continue to be inadequately maintained and hazardous. Old mills or small operations are most likely not to have proper ventilating systems and sanitary facilities. Workers have frequently left the mines in contaminated clothing when they lacked adequate washrooms and locker facilities (251). Moreover, smelting operations may be quite primitive. Since ore decomposes into metallic mercury at low heat, 338° to 420°C, it can be most simply and cheaply recovered by heating until the mercury vaporizes and can be collected in condensers. In some cases old furnaces and wooden containers lost so much mercury that the ground under them was worth "mining" to recover it. Fumes from these operations pose a much greater hazard than mining where the ore is not of high enough grade to exhaust mercurial vapors.

A mining operation in Sonoma County, California, was forced to close in the 1940s because a malfunction in the vapor-control mechanism of the furnace released enough fumes in 2 weeks to cause toxic symptoms among the workers. Relining the furnaces and welding the condensers and retorts also exposes repairmen to more intense concentrations of vapors for short periods of time. A California mechanic's lips became so swollen that he could not speak or chew, and even swallowing liquids was painful after he worked 1 week replacing fire bricks in an ore furnace (252).

In 1965 approximately 170,000 lb of mercury were released into the atmosphere as the element was mined, smelted, and refined (253). Although stack losses are not expected to exceed 2 or 3% unless the operation is highly inefficient, even a 3% loss could mean that over 50,000 lb of mercury were emitted into the atmosphere of the United States in 1966 by smelting alone (254).

10

OBSOLETE APPLICATIONS

MANUFACTURING FELT HATS

More myths and legends have evolved about the manufacture of felt with mercury than any other process since alchemists sought the philosophers' stone to turn the liquid element into gold. Whether the term "mad as a hatter" derived from the many hatmakers stricken by mercurialism or from Lewis Carroll's *Alice's Adventures in Wonderland*, this author immortalized the disjointed perception of reality displayed by the Mad Hatter as he takes tea with the March Hare (255). An alternate theory of the origin of the term "mad as a hatter" is that an old French reference to an angry viper may have been translated "mad as an adder" and then been given an initial "H" in the Cockney dialect (236). A Chesham hatmaker named Roger Clap also has been called the original mad hatter, because this eccentric soul was thought demented when he insisted on giving all his worldly possessions to the poor in 1650 (256). No one has reported that mercury influenced his generosity, but "mad as a hatter" and the "Danbury shakes" have long been colloquial terms to designate the mental and physical symptoms of inorganic mercury poisoning in the hat-making industry.

According to legend, St. Clement, the Roman patron saint of hatters, accidentally discovered felt while on a pilgrimage to Jerusalem. After lining his sandals with camels' hair to ease his feet, the interaction of heat, pressure, and perspiration produced a sheet of felt under his soles. In the

129

Orient this primitive method of felting was continued well into modern times (237). An alternate tale contends that felting originated in Turkey where camels' hair was made into tents, and the felting was accelerated by adding camels' urine to soften the hair. The Crusaders brought the art back to Western Europe where French hatmakers felted the fur of rabbits, rats, and other animals. Lacking camels, the workers urinated into a bucket and carefully saved the contents to soften the animals' hair. One workman's urine produced a more luxuriant felt after he was treated for syphilis, and then mercury was identified as the magic felting ingredient (236). Workmen were relieved of the obligation to produce the felting liquid when a mercuric nitrate solution was devised to soften the stiff animal hairs. This process, called "carrotting" because the mercury turned white rabbits' fur a reddish brown, was a carefully guarded trade secret. The liquid was called "le secret," and French hatmakers maintained a monopoly on the technique called "secretage" until the Edict of Nantes was revoked in 1685 (257). Then the Huguenots took the secret with them when they emigrated to England, and the center for fine hatmaking shifted to that country for 40 years. A Frenchman subsequently carried the method back to France, and mercurialism also accompanied the hatmaking industry to the colonies whenever that industry was introduced (236).

Initially, Europeans conducted the three major phases of hat manufacturing: fur cutting, "making," and finishing in small shops or private homes. Belgian hatmakers long maintained a reputation for producing fine felt hats, because they plucked out the long hairs by hand. These coarser hairs were suitable for upholstery fabric but not soft hats. Consequently, whole families were afflicted with mercurialism when they performed this close hand labor. In the American colonies mercury exposure was complicated, because all phases of hat manufacturing from raw skin to finished product were frequently combined under one roof. As machines replaced hand labor, workers were still exposed to mercury vapors in the poorly ventilated workrooms.

For the first step, carrotting, French hat manufacturers devised ways to dress the fur side of the skin so the grease was removed and the felting properties increased before the mercuric nitrate solution was applied. Then the carrotting solution was scrubbed into the animal's skin by hand with a brush or by machine carrotting with the skin placed between rollers. Hand scrubbing caused more close exposure to mercury, but workers still removed the wet skins from the machines by hand. Furs that were carrotted but not felted immediately were dried and tied up in bundles to be stored. Then they had to be dampened again to soften the hairs for cutting. Both the steam chamber and the storage room often concentrated high levels of mercury vapor.

In the next step the fur could be cut from the animal's skin so skillfully that it passed into a tray looking as if it were still attached. Then a "locker" trimmed the outer edges and compressed the fur into a bag ready to be turned over to the hat manufacturer (258). He would mix and blow the various furs to blend them together and then measure out sufficient quantities for individual hats. Each portion was then spread on a cone, immersed in water, and passed into a hardener to be shrunk. In hand processing the cones of felt were dried with towels. Personnel called "wetters down" shrank and sized the hat bodies by separating and adjusting them for uniform shrinkage to the desired dimensions. Dyes were added as an intermediate step, and the hats were shaped and smoothed last, a process called "pouncing" (258–259).

Although wool felt hat bodies could be produced without mercury, for centuries other animal fur felt was thought superior while such popular felt headgear as the stovepipe, ten gallon, military, and business man's snap brim all held their fashion sway. Mercury remained the standard carrotting ingredient for centuries, although as early as 1765 a French workman named Mathieu returned to Paris from England with a substitute that was considered better than current French procedures. The basic solution was aqua fortas (nitric acid) diluted with water to which lard and sometimes honey were added (257). In 1817 Guichardier also tried sulfuric acid, but it did not carrott as effectively as mercury (260). In 1882 molasses and nitrous acid were combined in a felting solution. Other formulations contained varying strengths of iron nitrate, zinc nitrate, zinc chloride, and nitrous acid. Prior to 1902 British patents were also registered for a number of carrotting solutions including: iron nitrate or sulfate; antimony sulfate; sodium nitrate or sulfate; potassium nitrate mixed with oxalic and sulfuric acids; tin dissolved in nitric and sulfuric acids; and a caustic soda solution (257). But mercury was still generally preferred.

The adoption of substitute carrotting fluids was more strongly encouraged after the French Academy of Medicine demonstrated the dangers of mercury in the hatting industry in 1869. A steady stream of testimony appeared in that country before a very detailed law was passed in 1898 to protect hatmakers from mercurialism (260). Youths under 18 and women were prohibited from carrotting, plucking out the long hairs, or cutting fur where much dust accumulated. The law also required impermeable floors and clean walls in good condition. The carrotting fluid was to be mixed only at night in a separate room. Ventilating systems were required where the fur was carrotted, blown, and cut as well as in the drying and storage rooms. Better sanitary precautions, work clothing, washing facilities, and a separate lunchroom were also required. Moreover, workers in contact with carrotted fur were to be examined by a doctor every 3 months and transferred to jobs without exposure if signs of mercurialism were detected.

Despite these legal measures, 12 years later, in 1911 the 500- to 600-member hatters' syndicate in Paris acknowledged that 60% of the 250 hatters' furriers suffered some symptoms of mercury poisoning. One investigator noted that "nobody of course thinks of denying the shocking effects of the acid nitrate of mercury. Everyone knows it is a poison and recognizes the damage it causes" (260). After an International Labour Organization Conference in 1919, the French accepted less harmful mercury substitutes as the Russians had after their revolution in 1917 because of suffering among families engaged in the manufacture of felt hats (237).

Although British doctors were also supposed to notify health authorities of serious cases of mercury poisoning after 1899, the workers were afraid they might lose their jobs if they sought medical attention, as a Guy's Hospital investigator observed in 1901:

These people I have seen who are still employed at the factory have begged me not to mention to their employer that I have seen them, as they would be instantly dismissed. Two cases not mentioned here, as soon as they found out what I had called for refused to tell me anything (261).

Employers exerted pressure by countering that routine medical examinations would interfere with the workers' liberty, incur a loss of time, reduce production, and, therefore, lead to the workers' suspension and loss of wages. The manufacturers contended that hatmaking was a voluntary hazard like working in coal mines. But those who favored medical surveillance argued that coal mining hazards were accidental, whereas industrial poisoning could be prevented. With better medical surveillance by

1918 only 46 of the 208 cases of mercurialism in England were in the hatters' trade. None were reported after 1921, although small hatmaking firms in Manchester used mercury carrotting fluids as late as 1966. Then larger firms were apt to monitor vapors, whereas the small ones still subjected employees to more exposure.

Unlike their British and French counterparts, the Americans did not pass federal industrial hygiene regulations to protect felt hat manufacturing employees until 1941. Although some states adopted health regulations, two extensive Public Health Service reports were published to document the excessive exposure to mercury before federal action was taken. Hat manufacturing had been introduced to the American colonies in 1662 when the Virginia Assembly offered a reward of 10 pounds of tobacco for every good felt hat produced from the fur of native animals (260). The processes devised by their European predecessors were continued in the hatmaking centers that subsequently developed on the east coast in Philadelphia, Orange, Newark, New York City, Brooklyn, Yonkers, Bethel, and Danbury, Connecticut, the town for which mercurial tremors were named "the Danbury shakes" (258).

In the winter of 1858–1859, more than 100 cases of mercury poisoning were identified among hatmakers in Orange, New Jersey, and conditions became worse during the War Between the States when orders for large quantities of hats were met by using strongly carrotted fur and shoddy, a fibrous composition of shredded woolen rags. Subsequently, despite the repeated documentation of mercurialism among workers engaged in hat manufacturing, no legislation was passed to regulate conditions before 1941, because industry spokesmen contended that the situation had improved greatly since the Civil War. In 1878 the 1546 hatmakers in Orange, New Jersey, were still exposed to seriously deficient industrial hygiene, partly because of the strongly carrotted shoddy (259). In some instances "all hands in the shop within a few days were rendered unfit for work or had their health impaired" (237). Blowers and shippers had the highest percentage of casualties (262).

And conditions were no better in other eastern seaboard felting industries than they were in Orange, New Jersey. A survey in New York in 1912 indicated that 66.7% of the male carrotters showed some identifiable symptoms, whereas 40 of the 212 hatters' furriers were afflicted. A smaller proportion of the women were symptomatic, although they had a close exposure as they picked the long hairs out of fur (260). Workmen who drank more alcohol were thought to enhance the tremors or to "drown their sorrows" because of mercury-induced depression. Consequently, mercury tremors were sometimes confused with those from alcoholism or senility, but dissiminated sclerosis generates a coarser tremor (263).

Danbury, Connecticut, also had a high percentage of mercury sufferers when the hatmaking industry was surveyed in 1922. Out of 108 hatters and hatters' furriers, 53 exhibited at least two of the three classic physical symptoms.

The emotional reactions were most difficult to detect. Excessively shy workers tried to hide their condition rather than seek treatment as they became depressed and feared losing their jobs and their lives (237). Horrible nightmares reinforced the employees' desire to cling to their positions irrationally as the mercury intoxication increased. The nature of early warning intention tremors also reinforced the desire to continue with the job and the exposure. These tremors often occurred only when the worker was aware that he was being observed or when he was asked to perform a specific dexterity exercise that was not part of his regular routine. One worker was calm until an examiner talked to him. Then he began to tremble, and the uncontrollable movements became exaggerated when he was asked to line up wooden pegs on a board. The tremors disappeared when the worker rested or thought he was not being observed. Tremors were rare in employees with less than 6-months service, and the severity depended on the worker's individual sensitivity and sanitary habits as well as the duration of exposure. Moreover, the symptoms might be erratic.

Some workers were employed for many years before they experienced a reaction, and then it might disappear and not recur for months or even years. Consequently, many workers were so familiar with their jobs that they could continue even when they were incapable of learning a new occupation. At Orange, New Jersey, a worker had such severe "hatters' shakes" that he could not feed or dress himself. He steadied his gait while walking to work by pushing a baby carriage ahead of him. At the plant he had to be guided among the machines to his bench, but then he could carry on the work with which he had become familiar through years of practice. Thus he preferred to continue the exposure rather than risk losing his wages. When workers did end their exposure, the tremors usually disappeared but not always. Ten years after he quit work, an old Danbury hatter would still shake violently the moment someone spoke to him, although he had been sitting by a window quietly reading a paper just a moment before. When a third person was present, his daughter always had to feed and dress him (237).

Despite the protective legislation and substitute carrotting fluids utilized in other countries, chronic mercury poisoning still afflicted about 8% of the 529 fur cutters in five representative felt-hat factories surveyed in 1937 and 1941 (259). Among 534 hatters, 59 victims worked in areas with air containing from 0.2 to 0.5 mg Hg/m^3. As substantially more symptoms occurred when the ambient mercury exceeded 0.1 mg/m^3 (237), this level was then established as the maximum allowable in industry. Among fur cutters exposed to 0.25 mg Hg/m^3 of air or more, 22.8% had chronic mercurialism within from two to five years, whereas only 5.7% exposed to lesser vapor concentrations developed toxic symptoms in the same length of time (257). Among the approximately 30 to 50 plants operating in 1940, the five studied did not necessarily represent the highest mercury levels either (259). Most of the 32,855 people who worked with mercury in American industry manufactured felt hats, and between 1935 and 1941 up to 140 such operations were in existence at least part of the time.

The sophisticated techniques available to monitor mercury vapors in 1940 still left few clues to the variations in individual susceptibility that depend on age, sex, mercury sensitivity, and physical condition. Symptoms occurred in less than 6 months or over 25 years. Many workers had blackened teeth that gradually fell out in a predictable pattern. Those with healthy mouths often resisted mercurialism longer than those with gum irritations or unsanitary dental habits. Consequently, girls who usually took better care of their teeth than the men were less susceptible to mercurialism. But the nitric acid in the carrotting fluid irritated the lining of the workers' mouths, so more mercury was absorbed whenever nitric acid vapors were also present.

After the Public Health Survey was published in 1941, the Surgeon General recommended a ban on mercurial carrotting of hatters' fur as well as on importing fur processed with mercury. Connecticut adopted the recommendations as part of the state sanitary code and set a penalty for violations that reportedly ended the Danbury shakes. Of the 30 other states that passed similar regulations, New Jersey was the only one that still had a considerable felt-hat industry. The other three major hat-producing states—New York, Pennsylvania, and Massachusetts—continued to drag their feet. Nonetheless, better sanitary conditions, more mechanization, and substitute carrotting fluids decreased mercurialism in the hat-manufacturing industry by the end of World War II.

THE CASE OF THE BLUSHING DETECTIVE

Fingerprinting was added to the science of criminology after 1880 when Fauld introduced this system to identify criminals in England. He shared credit for the innovation with Sir William Herschel who brought back reports of how he had used fingerprints for 20 years in the Hooghli District of Bengal to safeguard against impersonation and repudiation of signatures. In the 1880s the distinctive skin furrows of the fingers were categorized for individual identification, and in the 1890s two systems of fingerprint classification evolved. By 1901 a fingerprint bureau was established at New Scotland Yard, and suspects' prints were taken by coating the fingers with a dark ink or powder and then pressing them on cards. Then the prints were developed by applying a fine powder; light on a dark background or dark on a light background. As methods of imprinting finger impressions for permanent records evolved, the system used to identify habitual criminals at Scotland Yard was also adopted by police departments throughout Britain and in many other countries (264–265).

Several mercury compounds could transplant the image of a fingerprint from the scene of a crime to a laboratory where it could be photographed (266). One of these was a gray powder (*hydrargyrum cum creta*) composed of two parts chalk and one part metallic mercury. Since it had a particular affinity for sweat-formed latent prints and produced a good photographic image, by 1949 approximately 90% of all latent prints in Great Britain were developed with gray powder.

But in that year health authorities checked members of the Constabulary force in Lancashire, England, because they had some unusual physical and emotional problems. A 45-year-old detective sergeant admitted that testifying in court cases had become a nightmare, because he became embarrassed and blushed easily. The sergeant and six other members of the police force had become increasingly irritable and uncomfortable when someone watched them work. In addition, seven of the total force of 32 men exhibited hand tremors. Three detectives also experienced tremors of the lips and tongue, and three more reported that their eyelids quivered. One man trembled so badly that he could not lift a full cup to his lips without spilling it. Some of the officers also noticed that their teeth were loosening.

The shy detective sergeant was trained in the fingerprint department in 1936 and had subsequently investigated about 300 crimes a year for three years. As he felt himself become a "nervous wreck," the detective gave up smoking, but this did not help. His nervousness diminished during the early years of World War II, but he did not associate this with the declining crime rate, although his symptoms became worse again in 1946 when more breaking-and-entering offenses occurred. Since all burglary suspects were routinely fingerprinted, the detective's work load increased with the crime rate, although his symptoms were not as severe as they had been previously because he had more men to help him. The detective did not suspect that his illness was job related until after health authorities ascertained that policemen with the most severe symptoms also did most of the fingerprinting (267).

As police work had become more specialized, some detectives spent much of their time at the central headquarters where they developed and photographed latent prints. Other specialists were stationed throughout the country where they could be called on to collect fingerprints at the scenes of crimes. Since men in both occupations were poisoned, it was possible to be a victim without approaching the scene of the crime. However, exposure was greatest when the fingerprints were collected. The detectives liberally applied the dry, finely divided powder with a well-saturated brush. Then they often cleaned the brush by pounding it against a convenient door post. The dust would fly about and settle around the room where it could be inhaled and absorbed through the skin. Gray powder was being "dusted" to fight crime at the rate of about 5 lb a month while the men brought their poison to work with them.

The diagnosis was confirmed after 31 of the 32 men on the force submitted to 24-hour urinary mercury analyses. Then the extent of their exposure was determined by checking the record of crimes that each man investigated. Detectives with symptoms of illness had normally spent from 250 to 460 hours at this occupation each year, although one victim had only 150-hours exposure. Among 11 men who averaged 250 hours or more per year either collecting or developing fingerprints, seven had severe symptoms. The other four who were asymptomatic probably were less sensitive to mercury, because they all had used the gray powder since 1936. When the cause of illness was identified, the officers recalled unusual behavior among their colleagues in the past which suggested that they also suffered from mercurialism. The officer in charge of the bureau recalled one colleague with the somewhat eccentric habit of not being able to tolerate being watched while he worked (267).

To prevent further mercury intoxication, rubber gloves and masks were considered but ruled out. Such devices would have added to the detectives' mystique, but their protection was inadequate. Therefore, mercury compounds were forbidden and a less toxic substitute was sought (268). In the meantime substitute compounds that had previously been replaced by the more effective gray powder were again utilized to take fingerprints. These included: aluminum, willow charcoal, acacia black, lamp-black, powdered graphite, and white lead. Then other police forces in England as well as the American Federal Bureau of Investigation were alerted to be on the lookout for blushing detectives.

11

MODERN INDUSTRIES

MECHANICAL EQUIPMENT

Elemental mercury has had an everexpanding range of applications based on such physicochemical properties as its uniform volume expansion, liquidity at room temperature, electrical conductivity, high density, low vapor pressure, ability to alloy with almost all common metals other than iron or platinum, and the fact that it can easily be frozen or vaporized. Industries utilize the element as a solid, liquid, or gas, and some functions depend on the transition from one state to another in manufacturing processes that range from the humble to the exotic.

Liquid quicksilver is one of 40 elements with a specific gravity greater than five. It weighs about 13.6 times as much as water, and a person can sit on a vat of the heavy liquid without sinking. The shear density of the element has been a factor in a number of applications. As early as 884 A.D. Abû-'L-Gesh Kumârawêyh, the Kalafan of Egypt, reportedly slept on an air bed that was laid on a lake of quicksilver nearly 100 feet square (269). More recently, a pool of over 22,000 lb of mercury was used as a frictionless support on which to float the optical assembly of a telescope (220). The heavy liquid has also served as ballast in the emergency flotation system of research and rescue submersibles. In an emergency one of these vessels could be trimmed by jettisoning up to 2500 lb of mercury into the water (270).

Bowling balls will also float on mercury. When they are made of rubber in two parts, a core and a veneer cover, a little "top weight" is sometimes incorporated in half of the core which is constructed of solid rubber and rubber mixed with a light material. The heavy side is then lost track of when the cover is added and the finished ball is ground to the specified dimensions. Therefore, floating the finished ball in a vessel of mercury causes the heavier side to rotate to the bottom. Then the point is marked, and the finger holes arc drilled in their appropriate positions (271).

Swallowing elemental mercury has caused very few medical problems. In fact, up to 2 lb have been swallowed without ill effect (272). Thus for centuries patients were given elemental mercury on the theory that the element would pass down the gastrointestinal tract in a lump and push any obstructions in its path (273). When it was discovered that the mercury actually could break into many small globules during passage through the body, the Canton and Miller–Abbott tubes were devised to hold the element together. Doctors then filled a balloon with about 25 g of mercury and inserted it into the end of a tube where the mercury forced its way through the intestinal tract but could not escape into the body. Afterwards, the element could either be reclaimed or discarded with the tube (274).

Because of its density, the element was also used as an anaerobic seal and mixing agent for a cardiac catherization syringe to draw blood samples from the human heart. If the attendant raised the syringe higher than the patient's body, the mercury could flow into his bloodstream and obstruct the arteries or veins. While large globules formed in the left ventricle of the heart, the diffusion could extend throughout the body. After one patient died and others who had undergone blood-gas analyses were found to have globules of metallic mercury in their bodies or symptoms of mercurialism, a stopcock was added to these syringes to prevent the loss of mercury.

The element was thought to occur only as a liquid or gas until two Russian scientists inadvertently observed that it solidified at approximately $-40°F$ outside their laboratory during the intensely cold winter of 1759–1760 in St. Petersburg (275). In its solid state the white metal is as soft as lead. A liquid mercury-cadmium amalgam can be poured into rubber castings that are frozen to make plaster casts for jewelry. After the casts are shaped, the amalgam can be heated to a liquid, poured out, and recovered while the plaster molds are filled with gold or alloys. Thus in a small jewelry-manufacturing operation in Wellington, New Zealand, workers were exposed to mercury vapors in rooms where the jewelry molds were formed, the plaster cast, and the amalgam heated (276).

The heavy silver liquid is most familiar in slippery globules that roll on the floor when a thermometer or barometer breaks. Mercury has become the standard measuring component for temperature and pressure gauges since Torricelli discovered the barometer in 1643 and Fahrenheit the thermometer in 1720 because of its great density, 13.6 g/ml at 20°C, and an almost constant thermal coefficient of expansion over a temperature range from 356.58°C (673.84°F) to −38.87°C (−37.97°F) where it freezes. The temperature range of the liquid can be extended to −60°C if a mercury alloy is formed with 8.5% thallium.

Thermometers are easily broken because they are generally made of glass, a substance that mercury does not wet. They usually contain only 0.4 to 1.0 gram of mercury, and individuals risk greater injury from swallowing the broken glass if they bite down on the stem of a thermometer than from the mercury itself (277). As with making plaster casts for jewelry, exposure to mercury vapors is much greater in the manufacture of thermometers than in their use. British workers developed such symptoms of inorganic mercury poisoning as nervousness and tremors as late as 1946, because much of the manufacturing was still done by hand. The glass was blown and the tubes heated to remove pockets of air before they were filled by inversion over bowls of mercury. On other occasions this ongoing exposure has resulted in more serious occupationally induced illnesses. In 1970 a 59-year-old man died without showing the usual symptoms of poisoning after filling thermometers with mercury for only 6 months. The attending physicians concluded that some degree of bone marrow suppression may occur with a high body concentration of mercury (278). And since the Becton–Dickinson Company of Rutherford, New Jersey, moved its Nebraska thermometer-manufacturing plant to Juncos, Puerto Rico, mercury levels have risen in the Valenciano River and broken thermometers can be found along the banks. In addition, the employees are meeting steadily more opposition from officials of the State Insurance Fund when they claim disability for mercury-related illnesses. Compensation was faster before 1969 when the state health officials concluded that the company had come into compliance with acceptable health standards by taking new safety precautions. Thus although the mercury vapor levels remain high enough to cripple the workers, they now wait from 4 to 8 years to have their claims recognized and be compensated.

Even resident animals suffer. The levels of vapors were so elevated at one thermometer factory that the problem was revealed to a veterinarian when the owner consulted him on how to rid the building of rats. He was reluctant to sacrifice another cat, because three had died the previous year after suffering inflamed gums, loose teeth, and paralysis. The rats were similarly afflicted and could be seen dragging themselves to their holes (279). Conditions improved for the employees as well as the animal inhabitants after the health authorities were notified.

When thermometers are broken, the small quantity of mercury that escapes into the floor is not apt to elevate vapor levels in private homes, although this is a possibility in hospitals or other laboratories where breakage is more common and other mercury-containing equipment is also used. Canadians break in the range of 3 million thermometers each year for a loss of approximately 14,000 lb of mercury (280). For instance, the University of Saskatchewan annually breaks about 2000 thermometers containing from 5 to 10 lb of mercury. Most of it is in clinical thermometers used by the Department of Veterinary Medicine, and the element has generally been flushed down the drain (274).

Whereas these minor losses are relatively insignificant, the element is far more likely to concentrate in hospitals, especially teaching facilities and college or university physics and chemistry laboratories, where spills from equipment accumulate in benches, cracks in the floor, and down the backs of mopboards (281). The quantity lost depends on the teaching tools and research in progress as well as the competency and sanitary practices of the technicians and the age and condition of the building and equipment.

Sophisticated electronic instruments may be economically feasible to replace mercury barometers and thermometers in some instances where the high rate of breakage and spills create a hazard, such as when hospital technicians fill thermometers to take gas samples without using a stopper. Then the mercury overflows on hands and clothing to come to rest in cracks in the floor. Mercury has also been eliminated as a catalyst where automatic analyzers have replaced the older manual methods for determining carbon, hydrogen, and nitrogen as well as carbon dioxide, electrolytes, and other constituents in blood.

Despite these advances, technically obsolete mercury-containing equipment may be retained for teaching certain principles that cannot be demonstrated visually with the electronic equipment. Until the 1940s when an oil was developed that did not break down under stress, direct blood pressure readings were demonstrated with small mercury manometers that would occasionally generate enough pressure to blow mercury droplets around the room. These and other experiments are not apt to release much mercury individually. Thus vapor levels are generally relatively low in high school chemistry and physics classrooms, although mercury is used in electroplating experiments as well as to demonstrate such physical properties as density, surface tension, capillary action, electrical conductivity, thermal expansion, and barometric pressure (281). Mercuric chloride is often used to demonstrate a precipitating agent, and mercuric oxide may be heated to illustrate the decomposition of a compound. In addition, students continue such traditional, unauthorized experiments as coating pennies and rings with mercury.

Although these occasional demonstrations are unlikely to cause mercury poisoning, some teachers in Saskatchewan eliminated these common experiments to prevent exposure, and they encouraged students to bring the mercury from home chemistry sets to school to dispose of it safely (274). Increasing student awareness is warranted because occasional spills of elemental mercury in private homes have been sufficient to raise the vapor levels and develop inorganic mercury poisoning in the family. One child stole some mercury from school and was ill for several years after playing with the element and spilling it in his bedroom.

The wide acceptance of mercury in mechanical equipment has made it an invisible hazard in laboratories, because it is so common that negative side effects are unexpected. At the University of Michigan a professor suffered more exposure in his office than his graduate students did in the laboratory. Although they worked near a large open container of mercury, less of it vaporized from this one source than from the many particles that were tracked into the office.

In other laboratories the operation of specific equipment may expose only one or a few individuals to mercurialism. In 1923 the Bureau of Standards operated two induction furnaces that leaked mercury vapors in metallurgical chemistry laboratories, especially when they had to be cleaned. After mercurialism was diagnosed in several workers, a new discharge gap and forced draft ventilating hood were mounted on the furnace to prevent vapors from leaking out (282). Such leaks also raised mercury vapors from 130 to 180 times acceptable levels in several American petroleum laboratories surveyed in 1939, but these high-induction furnaces continued to be used without proper ventilating equipment. In 1954 the exposure of a 28-year-old chemistry research assistant in a university laboratory demonstrated the insidious effects of this gradual mercury intoxication. The assistant was the only one who opened the unvented furnace for about 2 minutes twice a day to remove a sample. But that short exposure was enough to cause him to undergo significant psychic changes. When his condition deteriorated from fatigue and intellectual sluggishness to pain and depression, he was no longer able to visualize organic chemical formulas or to comprehend his course of study. Consequently, until his condition was diagnosed, he usually remained quietly by himself except for occasional irritable outbursts at his family (283).

Mercury poisoning may also have been a hazard for other scientists whose affliction went unrecognized in the past. Two American physicists have concluded that Sir Isaac Newton had a classic case of mercury poisoning rather than the dementia his biographer attributed to a psychic trauma caused by his mother's death when Newton was 50 years old. He worked in closed rooms where mercury vapors could easily exceed modern safety standards by a factor of 1000 or more, and he preferred to live in crimson rooms that could have been painted with a formulation containing mercuric sulfide. In addition to the likely exposure, the diagnosis was supported by analyzing examples of Newton's tremulous handwriting (284).

By far the most famous case of inorganic poisoning was recorded by Professor Alfred Stock (1876–1946) whose 1934 studies of mercury in food and the environment are still widely cited (285). Stock began to work with mercury in his home laboratory during his school days. He developed early symptoms of mercury poisoning such as headaches, dizzy spells, numbness, and inflamed mucous membranes after 1900 when he assembled a small mercury trough in an unventilated room. The symptoms decreased when Stock was not conducting experiments and intensified when he resumed them. As his upper respiratory tract was afflicted, his memory and hearing also continued to deteriorate until by 1924 he was unable to deliver a lecture without extensive notes, and he forgot telephone numbers between the directory and the phone. Nonetheless, Dr. Stock did not associate his illness with exposure to mercury until his assistant, Dr. Wolfhart Siecke, developed the classic sore mouth and loose teeth. Stock then concluded that he had not contracted oral symptoms sooner because of the exceptional care he had taken of his teeth since boyhood (286–287).

After recognizing his susceptibility to mercurialism, Stock painstakingly outfitted his new laboratory with seamless flooring and took other precautionary measures as well as conducting experiments to determine the effects of mercury. He devised a new analytical technique that proved mercury was more dangerous as the vapors passed into the respiratory system than when the element was consumed and passed from the digestive tract into the circulatory system. Stock demonstrated this with animal experiments and also by injecting a dilute mercuric chloride solution into the back of his own nose. The typical symptoms of mercury poisoning were evoked "doubtless to his discomfort as a victim as well as his satisfaction as a scientist" (287). Because of the mercury poisoning that had previously been diagnosed as arthritis, Stock increasingly curtailed his activities and was finally forced to retire in 1936. His detailed account of his worsening health was published to "spare them (other scientists) the terrible experience that spoiled a large part of my life" (288). His research into the lives of other scientists such as Faraday and Pascal convinced Stock that they had also unknowingly suffered chronic mercury poisoning (288–291). At his death in 1946, Stock left an extraordinary scientific record both of his research on inorganic chemistry and the health hazards associated with mercury poisoning (287). And as early as 1928 Stock's experience led another scientist to describe his suffering from mercury poisoning.

During the coldest days when one was forced to heat a lot, I had placed on a stove which was heated twice a day; an open dish with approximately 20 lb of mercury in order to dry it well. It is known that mercury requires a very high temperature in order to vaporize and that it appears to condense again out of the air very quickly when the temperature is decreased. The temperature of the air in the room where I had the mercury on the stove and where I would spend almost the whole day was usually between 14–18°R (64–73°F). After some days I felt ill although I paid no attention. The discomfort became greater; I had a pain in the throat; my gums began to hurt and became inflamed; and I salivated copiously. Because I did not think of the mercury on the stove, I attributed this illness to having broken a flask in which I had distilled mercury on a daily basis. On the same day that I began to have the salivation symptoms, two other people felt the same symptoms although they had continually been in the same room and had never set foot in the laboratory where the flask was broken. They also began to salivate the next day. Now I searched carefully for the cause. I could find nothing except air contamination from the mercury that had been placed on the stove and had escaped because of the warmth (291).

ELECTRICAL EQUIPMENT: AN EXPANDING FIELD

Since Thomas Edison introduced the first successful incandescent lamp in 1879, electrical power gave mercury a new outlet with a continually expanding range of applications (292). In the 1960s the electrical apparatus and chloralkali industries were the fastest growing markets for mercury. By 1968 electrical apparatuses required 26.8% of the total mercury utilized industrially and in 1969 surpassed the demand of the chloralkali industry, the second largest mercury consumer, by nearly 5000 flasks. This level of growth had been projected for 1974 only a year earlier. By 1971 nearly a third of the total 52,475 flasks went to meet electrical demands (293). Then electrical equipment, the chloralkali industry, and the much smaller industrial and control instruments manufacturers accounted for two-thirds or 2,589,396 lb of the total mercury utilized by industries in the United States and Canada (293–294). A 22,000-flask consumption rate was projected for 1979, the centennial year of Edison's invention, but this is also likely to be exceeded because of the greatly expanding demand (270). Most of the mercury goes to manufacture dry cell batteries and new electrical equipment as well as to calibrate apparatuses in the glass industry.

Mercury is an excellent conductor of electricity, because it offers little resistance to the flow of electrons. Its electronic applications include phanotron, thyratron, and mercury-pool cathode tubes as well as high-frequency applications in radar, radio, and welding. Mercury has also graduated to the space age in liquid flywheels and gyroscopes to stabilize spacecraft as well as in fuel cells, ion engines for rockets, and turboelectric power systems (221). Whereas mercury-vapor-powered turbines proved to be unsuccessful in 1923 because of the cost, weight, and corrosion problems, 30 years later when mercury was again tested to determine if electric energy and fissionable material could be produced from experimental reactors, the Rankin cycle turboalternator was developed as an auxiliary power conversion system for space flights. The high thermal conductivity also enables mercury to serve as a coolant in nuclear reactors, whereas the thermal neutron-capture cross section (360 barns) permits the element to absorb neutrons and shield atomic devices. Two lithium isotopes, numbers six and seven, can also be separated with mercury. Lithium six is made into lithium tritide, an important H-bomb ingredient. The more numerous, prosaic electrical applications include: lamps, motors, meters, oscillators, switches, and batteries. They depend on mercury's electrical conductivity as a vapor or a liquid, and their manufacture exposes workers to varying degrees of mercurialism depending on the sanitary conditions at the factories.

Direct-current meters were manufactured in small electrical shops at the turn of the century. Workers both inhaled and absorbed mercury as they washed the meters in an acidic solution of mercuric nitrate and then filled them by immersion in metallic mercury so that the armature revolved in a mercury bath positioned between the poles of a magnet (295) (see Figure 19A). At mid-century workers were still suffering from mercurialism as they repaired these direct-current meters in small London shops with high temperatures and a lack of ventilation. Working with the element more than 5 years often imparted a brown coloring to their eyes from mercury particles while their hands developed the conventional tremors. Figure 19B illustrates a workman's tremors as he attempted to draw a picture of a direct-current meter when he was admitted to the hospital in 1949, and Figure 19C shows that the tremors lessened considerably in the 12 months after his exposure ended (296).

Mercury-arc rectifiers were developed to convert alternating electrical current into direct current. The basic principle is that an electric arc conveys the current between the conductors by means of gases. Mercury vapor is an effective conductor between the anode and cathode, because it is liquid at normal temperatures, it vaporizes freely, and it conducts electricity well (292). Glass-bulb rectifiers provide up to 1000 kilowatts of electricity, whereas metal-tank rectifiers can be built to any size needed to supply the high-voltage requirements of industry.

Many of the hazards from mercury-arc rectifiers could be readily corrected. For instance, in their initial production mercury was poured into small rectifiers on top of a processing oven at temperatures up to 400°C. Any spills vaporized directly in front of the workmen. By developing removable containers with exhaust pipes, the mercury could be carried

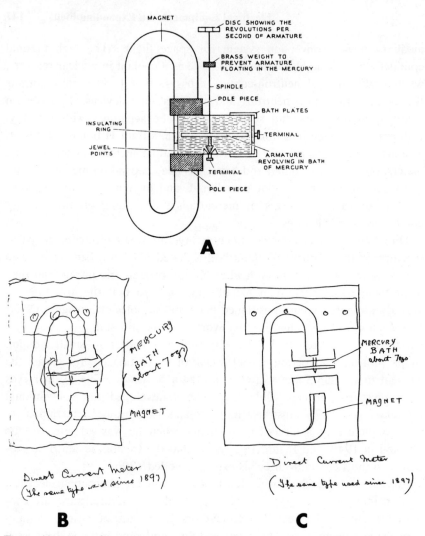

Figure 19. (A) An illustration of a mercury direct-current meter. (B) A sketch of a direct-current meter by a workman suffering from mercurialism on his admission to the hospital. (C) A sketch of a direct-current meter by the same workman 12 months after his removal from exposure to mercury. From *The Lancet*, **2**:856, 1951.

148

away from the oven before it was poured into the vessels. Vapor levels also were high in areas where the arc rectifiers were repaired, because the mercury sludge had to be removed from the vacuum envelope with a rotating brush. Workers were exposed to the vapors during this operation and from mercury that had collected on the floor. In one wire-brushing area approximately 10 lb of metallic mercury had seeped through cracks in a parquet block floor over a 5-year period. Vapor levels were lowered by replacing this floor with a smooth vinyl sheet that sloped to a ventilated mercury sump (297). Mercury rectifiers are also gradually being replaced by solid state rectifiers. In 1964 a 10% decline was anticipated by 1979. Nonetheless, mercury consumption was still expected to increase in the expanding relay industry (270).

Lights, switches, and wires all involve mercury and expose workers to inorganic mercury poisoning during their manufacture. Tungsten-molybdenum rod-and-wire combinations for electrical contacts and filaments are heated in liquid mercury to make the metal particles more cohesive. When an electric current is passed through this medium, the metals shrink and coil into a brittle rod that can be heated until it is malleable enough to roll into wire. In 1945 the exposure rate of workmen was correlated with the number of treatment processes operating at once and the amount of mercury that was spilled (297). Silent wall switches also turn lights on and off with a bead of mercury that conducts electricity between two metal contacts as it moves from one end of a glass tube to the other. In 1963 workers suffered a number of symptoms of mercurialism after pouring mercury into the glass bulbs in which the electrodes were sealed for these switches. One employee had such pronounced tremors that he discontinued such activities as playing darts, tinkering with watches, and even sending out Christmas cards because his writing had become so irregular. Another victim's hands shooks so severely that he could not hold a cup of tea. He also complained of fatigue and depression. A third employee was so emotionally unstable that he burst into tears during an interview, because he was disturbed by noise from an air hammer outside the consulting room (298).

By far the greatest numbers of electrical workers are exposed to inorganic mercury poisoning in the manufacture of mercury-vapor lights and Ruben batteries. Incandescent street lights, flood lights, neon signs, and industrial lighting are now often replaced by the more efficient mercury lamps. In 1968 the current 1200-flask requirement to fill these lights was predicted to double in the next 10 years, since no substitute was available (270). One factory alone purchased 15,000 to 20,000 lb of mercury annually for this purpose. Since an average of 200 lamps were broken in each factory unit per day, exposure to the vapors after the element escaped into the cracks in the cement floor became a point of contention between workers and management at the International Telephone and Telegraph Company in Lynn, Massachusetts, in 1972. The workers demanded rotation to jobs that did not expose them to mercury as a safety precaution (299).

The second wide application for mercury developed from one of the most significant electromechanical innovations of World War II. For many specialized uses, the Ruben or mercury battery has all but displaced the conventional dry cell battery since 1944 because of its longer life, smaller size, more stable voltage, and more efficient utilization of the active ingredients. The mercury battery also has a longer shelf life and can deliver the same ampere-hours service whether operated intermittently or continuously. Moreover, mercury dry-cell batteries can stand up under high temperatures and humidity. Consequently, they are the power source for most hearing aids, portable radios, miniature flashlights, walkie-talkies, Geiger counters, and more sophisticated equipment in electronic measuring devices, digital computers, guided missiles, and spacecraft (221,300).

Workers have had limited exposure to mercury in the manufacture of these batteries, because the equipment has been automated, and careful industrial sanitation has often been maintained in the plants, although mercury vapors circulated through at least one factory's central air conditioning system. Conditions can be controlled while a mixture of mercuric oxide and graphite is pelleted and placed in the battery containers as a cathode. But working conditions have been less carefully monitored in small operations to reclaim the valuable components of these batteries. Twenty-nine workers in one small shop processed 55 kg of a zinc-mercury amalgam from old airplane batteries to obtain zinc oxide in 1963, and the plant closed before mercurialism was suspected. After a former worker's tremors were observed in a routine medical examination, six out of the ten employees who had the greatest exposure were hospitalized (301).

Instead of reclaiming the components, old batteries are more often dumped in landfills where the casings eventually corrode and release their contents to leach into the ground. Some batteries are also burned, and the mercury vaporizes into the atmosphere (302). Since at least 500 tons of mercury are consumed annually to manufacture both alkaline-energy and mercury-cell batteries, they are by no means an insignificant source of mercury to the environment.

Mercury contamination and consequent poisoning can be prevented only if the employees recognize the hazard and take the proper precautions. However, the symptoms can be so general and develop so gradually that even wholesale mercury poisoning has continued undiagnosed for several years. Mercury poisoning had a significant impact on the physical and mental well-being of workers who produced carbon brushes for automotive generators in a small, poorly ventilated factory until the problem was discovered during a health investigation in 1953.

Despite the fact that women who were hired from the town and surrounding area often experienced bleeding gums and a metallic taste within a week after starting work, the specific relationship between employment at the factory and the physical and emotional disabilities was not pinpointed before the investigation. Nonetheless, absenteeism was frequent among women employees who often consulted local physicians and dentists about their illnesses. New employees learned to expect a sore mouth, and the company maintained a supply of mouthwash for their use. The women diagnosed their condition as trenchmouth, although they did not pass it on to their families. It was common only among employees and became more exaggerated in the winter, probably because of poor ventilation. As several women developed a blue line on their gums, a local dentist suspected that his patients' unhealthy mouth condition might be occupationally related at least 3 years before the health investigation. However, officials denied that any toxic substance was being used, and no further action was taken.

The women tamped a copper amalgam compound into small brushes by hand and into larger ones with semiautomatic machines to cement a copper tail wire into a hole in a carbon block. Then the brushes were baked in ovens, and the free end of the tail wire was either soldered by hand or crimped with clips. A pail of the tamping compound stood open, and substantial quantities of it spilled on the floor and benches where the women worked, ate, and smoked. They were exposed to the 22%-mercury tamping compound through inhalation, absorption, and ingestion as it covered their hands, faces, and clothing. The women also carried the contaminant home on their work clothing, because the plant had no facilities to shower or change.

The workers complained of nausea, vomiting, headaches, blurred and restricted vision, and occasional fainting spells. One Friday morning during the investigation, six employees collapsed into unconsciousness, and the plant manager administered oxygen. The handy oxygen tank suggested that such fainting spells were not unusual. Many women also experienced chest and muscle soreness and lost their coordination and acuity. Consequently, their work output decreased, and they wasted more material. One machine tamper caught her fingers in her machine several times. Some of the women's handwriting became so illegible that even their signatures could not be read, and one 60-year-old woman was released after 5 years' employment because her hands became so limp that she couldn't grip things.

As some individuals became passive, family members noticed that their skin also developed a grayish pallor. Some reported a greenish perspiration that discolored their clothing and bed linen, and husbands complained that the women had a peculiar odor when they came home. Nonetheless, employees worked until they became ill, took time off to recuperate, and then doggedly, "almost hypnotically," returned to work. They paid their own medical expenses, because the illnesses were not believed to be job related and because this factory was one of the few employers in the community.

Some women also experienced emotional disorders that made it impossible to evaluate the impact of the gradual changes on themselves or their families. Some of them began to neglect family duties. For instance, one woman forgot such routine things as buying groceries and starting to cook. Social life became negligible as they developed a feeling of semi-stuporous detachment from their surroundings that gradually led to a sense of isolation at home, work, and in the community. One previously cheerful woman either became jittery and snapped at everyone or withdrew and sat quietly. When someone talked to her, she seemed to ignore them, although she later said she heard them talking but was just too tired to answer. Although she was drowsy all the time, this women had nightmares, and her husband thought she was losing her mind. Another woman reported that

> I got so grouchy and nervous I would cry at nothing and I never went anywhere at night with my husband. At first I would go to bed and sleep so sound I would get up tired the next morning but later I got

so my sleep was so restless that I often would awaken so suddenly and have a fluttery feeling like I was scared or floating in space (303).

The women's lives also had a nightmarish quality. They lived on such an emotional fringe that they were easily irritated and would cry with little provocation. When the six employees fainted, the others became hysterical. Even after the cause was discovered, attempts to improve working conditions had to be handled delicately, because the women were emotionally unstable and reluctant to accept criticism.

As they were educated to take precautions such as wearing protective clothing, keeping the work area clean, and eating and washing in separate facilities, the gray pallor left the employees' faces, and they became more alert and interested in their surroundings. Some of the women were subsequently awarded workman's compensation for their job-related illnesses, but others had their claims denied because the statutory time limit had lapsed. Nor could any correlation be confirmed between longer range problems and mercurialism, such as when three of the women miscarried a year after being severely affected, and two long-married women, also severely afflicted, separated from their husbands.

The cause of illness probably was not detected more readily because the symptoms and mercury blood levels varied so widely. In 1953 the plant listed 78 current employees of whom only 36 were on the job and 49 had worked less than six months. Although the turnover was generally rapid, some of the women had worked at the plant since 1946 (303). Two women with long work histories had no symptoms despite their elevated blood mercury, whereas others were symptomatic in a much shorter time. One woman exposed only three days had 215 ppb mercury in her blood. Generally, mercury blood levels ranged from 5 to 600 ppb, whereas urinary levels ranged from 10 to 3740 ppb. This long-term tragedy demonstrated the need for better supervision of industrial hygiene since the workers were not able to recognize and cope with the problem themselves.

INDUSTRIAL INTERMEDIARIES

Although more mercury is consumed in electrical equipment than in the manufacture of chlorine and caustic soda, the use of mercury in chloralkali plants and as an intermediary in the manufacture of plastics and other products is the source of many point discharges to the waterways and atmosphere. The disaster at Minamata was triggered by methylmercury that was inadvertently produced and discharged into the bay during the manufacture of acetaldehyde. Nonetheless, the demand for mercury catalysts to prepare vinylchloride, urethane, anthraquinone derivatives, and other chemicals has risen steadily. Among these catalysts are mercuric chloride, oxide, sulfate, acetate, and phosphate salts.

As the list of plastic products has steadily extended, so has the demand for polyvinylchloride (PVC) and urethane. Although PVC was discovered in 1835, it was not manufactured commercially until 1927. Twenty years later, 100 million lb were being produced annually, and the worldwide output had expanded to nearly 3 billion lb a year by 1968. About 0.074 lb of mercury is consumed to manufacture 1000 lb of vinylchloride from acetylene. Thus 38,000 lb of mercury were required in 1968 to manufacture the most widely used plastic in the world. PVC is familiar to nearly everyone as phonograph records, printing plates, toothbrush handles, bottles, ice cube trays, pipe fittings, valve parts, and siding for buildings. Flexible PVC is made into raincoats, shower curtains, tablecloths, draperies, refrigerator bowls and covers, babies' plastic pants, auto seat covers, furniture upholstery, vinyl floor tile, tubing hose, weather stripping, wire, toys, buttons, and innumerable other applications as plastics have become the great imitators of wood, leather, and chrome.

Mercury has been an indispensable catalyst in the dye industry since about 1887. After Adoph von Bayer synthesized indigo dye in 1880, the Badishe Company spent nearly $5 million trying to convert the laboratory synthesis into a commercially feasible process. The most difficult stage in manufacturing the dye was to convert naphthalene into phthalic acid. A laboratory technician unwittingly solved the problem when he broke a thermometer in a container of molten napthalene. With the mercury catalyst the compound commenced to boil and converted into phthalic acid quickly and smoothly. Nor is any substitute available for mercuric oxide as a catalyst in sulfonated anthraquinone products. As mercury catalysts continue to grow in popularity for uses in photochemistry and other sulfonation products, Ziegler syntheses, polymers, and to manufacture fluorine chemicals and dyes, 45,600 lb of mercury were required in 1968, and the projected annual rate of production was at least 120,000 lb as these uses continued to expand.

Similarly, mercury-cell chloralkali plants projected a continued growth rate of up to 7% annually before they were confirmed as the point discharge sites of large deposits of elemental mercury in waterways where it was being methylated naturally and concentrated in the fish. By the time pressure was brought to bear on the American and Canadian chloralkali industries to decrease their discharges in 1970, a 5-mile-long layer of sediment with up to 1700 ppm of mercury coated the bottom of the St. Clair River downstream from the Dow Chemical Company at Sarnia, Ontario. Much more of the element had also been washed downstream to Lake St. Clair and had been dredged out and dumped into Lake Erie. Downstream from other plants the pattern was the same. Up to 1800 ppm of mercury were measured in the sediments near a chloralkali plant in Saskatoon, Saskatchewan (144,272). When the Dow Chemical Company of Canada at Sarnia, Ontario, released up to 200 lb per day, the mercury could be worth $450,000 a year at the record 1969 prices. The American chloralkali industry consumed as much as 3300 lb of mercury per day or 1.3 million lb per year, more than any other industry (125). Similarly, the Canadian chloralkali industry required 195,000 lb of replacement mercury annually, 20% of that nation's total industrial consumption (304). In 1970 eight Canadian chloralkali plants and eight pulp and paper mills consumed about 200,000 lb of mercury. Most of the plants were in Ontario and Quebec provinces on waterways that bordered the United States, so industrial contamination affected both countries (118).

Producing chlorine and caustic soda with mercury-cathode electrolytic cells had been suggested originally in 1883, and the first commercial operation began in 1894 at Oldbury, England, where H. Y. Castner erected a battery of mercury rocking cells. The Mathieson Alkali Company then purchased the American right to this process and erected the first plant in the United States at Saltville, Virginia, in 1895. The Canadians built their first chloralkali plant near Windsor, Ontario, in 1911. This diaphragm cell plant was converted to mercury cells after a purer caustic soda was required to produce viscose rayon in 1925 and transparent cellulose film in 1932 (305).

At the present time, companies often own "captive" chloralkali plants to supply their needs exclusively, because chlorine and caustic soda are basic ingredients in the production of numerous chemicals. For instance, seven of the eight Swedish chloralkali plants are owned by the cellulose industry. In the United States 50% of the chlorine is consumed in plastics, whereas the aluminum industry requires the most caustic soda, followed by glass, paper, petroleum, and detergents.

Caustic soda (lye or sodium hydroxide) and chlorine gas are basic chemical intermediaries in a number of industrial products, many of which ultimately are intended for consumer consumption. Chlorine is the active ingredient in household bleaching powders and liquids. Since it kills germs quickly and leaves no taste, chlorine also is the disinfectant in swimming pools, and by 1937 it was the sterilizing agent for over 75% of the drinking water in North America. Chlorine was credited with decreasing the death rate from typhoid fever to less than 5 per 100,000 population (306). It also is used to disinfect sewage and as an intermediary in the manufacture of dyes, insecticides, fire-extinguishing liquids, grease removers, and synthetic fibers. Chlorine or its derivatives also are added to newspaper, textiles, and medicines.

The companion product, caustic soda, is also used to manufacture soap, rayon, paper, medicines, and textiles. Dyes such as blue indigo and red alzarin as well as other organic compounds are also synthesized with caustic soda. And it is an ingredient in the production of resorcinol, which is used in skin lotions and to tan animal hides. Formates and oxalates are also manufactured with caustic soda. These are intermediates for refining the basic ingredients in plastics and other petroleum products. When produced with mercury cells, caustic soda may also introduce traces of mercury into the human food chain from processed foods. Sodium hydroxide is an aid to peel potatoes and fruits, to glaze pretzels, to neutralize sour cream butter, and as a reagent to refine vegetable oils and animal fats as well as to adjust the alkalinity in canned vegetables.

As the list of uses for chlorine and caustic soda lengthened after World War II, the chloralkali industry steadily expanded to a growth rate of 7¾% annually in the 1960s. From less than ½ million tons in 1935, chlorine production increased to a record 10,419,478 tons in 1972 (307). This growth rate was expected to decrease to 6% by 1975 as manufacturing needs were met, and in November 1971 the chlorine industry was operating at 85% of capacity. However, as the demand continued to rise for caustic soda that was less contaminated with chloride ions, more of it was produced with mercury cells. The percentage went from 4 in 1946 to 12 in 1956, 21 in 1963, and 30 in 1969, still less than Canadian mills where 60% of the chlorine and caustic soda were produced with mercury cells in 1972 (294,307). American mercury cells were projected to increase production to 38% by 1975 before the industry was pressured to reduce its mercury losses to the environment and before the 1974 recession (294,307–308). As the chapters on mercury contamination of the environment indicate, this is a much more serious problem overall than the workers' limited exposure to mercury vapors within the chloralkali industry.

Although mercury cells have been operating in the United States for more than 80 years, the long-term effect of chronic exposure to mercury vapors exhausted within the chloralkali plants has not been studied. However, most workers showed no discernible symptoms of poisoning after chronic exposure for more than 30 years, although hypersensitive individuals suffered acute symptoms more readily. For instance, two Swedish workers developed kidney injuries and albuminuria after less than a year's employment in 1952 (309). When 21 plants in the United States and Canada were checked in 1968, 14% of the 642 cell-room employees also were exposed to mercury concentrations that ranged from 0.001 to 2.64 mg/m^3 (254), and 27 workers had somewhat elevated blood and urine levels when mercury vapors ranged between 0.05 and 0.15 mg/m^3 of air (310).

Some effects of this exposure can be determined by comparing the workers with diaphragm-cell employees who are also exposed to chlorine gas but not to mercury. Mercury-cell workers have displayed more nonspecific symptoms of illness such as loss of weight and appetite as well as occasional dizziness. Those with greater exposure also had tremors and seemed to consume more alcohol. Although subjective symptoms were difficult to pinpoint, they also demonstrated more neuropsychiatric symptoms than their counterparts among diaphragm-cell employees. People in good physical health sometimes suffered from insomnia, shyness, and nervousness. Such nonspecific symptoms usually were not thought to be job related. However, after local newspapers described inorganic mercury poisoning, two workers at the Dryden Chemical Company in Dryden, Ontario, had physicians confirm that their headaches and irritability could be job related, although this was not proven.

Working conditions and individual sanitary habits still vary so much that it is difficult to predict the likely mercury intake and accumulation even when vapor levels are carefully monitored. They also may be inaccurate despite sophisticated equipment if low concentrations of chlorine gas in the cell rooms remove some mercury vapors from the immediate vicinity and lower the readings on the meters (311). Moreover, occasional leaks cause surges in vapor levels even where good ventilation is provided and spills are quickly washed into sump pumps, traps, or drains.

More mercury is exhausted when the supply is replenished or the area is cleaned. Workers normally absorb some mercury through their skin and inhale more if they do not wear respirators. A labor dispute temporarily closed the chloralkali facility of the General Aniline Film Corporation at Linden, New Jersey, in September 1971. Workers in the chlorine-caustic soda unit wanted to be rotated to minimize their exposure to mercury vapors, whereas management contended that the exposure was increased because the workers were reluctant to wear respirators (312). For whatever reason, employees in the chloralkali industry became more aware of potential hazards from exposure to mercurial vapors when public consciousness was raised concerning the vast quantities of the element that were being discharged into the waterways.

Since 1946 an estimated 13.8 million lb of mercury have been lost by the American and Canadian chloralkali industries (see Figure 20). Approximately 700 flasks (76 lb each) were required to replace losses in the chloralkali industry in 1947. As the demand increased steadily after that, about 4000 flasks were consumed annually between 1955 and 1959 and expanded to a record 20,720 flasks in 1969, despite the fact that new cells generally hold less mercury than the older models (313). Whereas the older units lost from 39 to 50 lb of mercury to produce 100 tons of chlorine, losses of 0.5 lb per ton were considered average and 0.4 lb were thought moderate. In 1970 techniques were quickly introduced to reduce losses from 0.5 lb per ton of chlorine to 0.5 lb of total production per day. A map with mercury-cell operations in North America is presented in Figure 21 (307, 314–315).

Mercury purchases declined to 15,011 flasks in 1970, although production was only slightly lower the following year. The output was still approximately 8000 tons of chlorine per day as 31 plants were still using mercury cells in April 1971, two less than when the federal crackdown began. But the Chlorine Institute reported that mercury losses had been reduced to an industrial average of only 2% of the January 1, 1970 rate 15 months later (312). As seven mercury cells discontinued operation altogether and another 13 converted to diaphragm cells, in 1973 mercury cells produced only 24.6% of the total chlorine. However, mercury losses

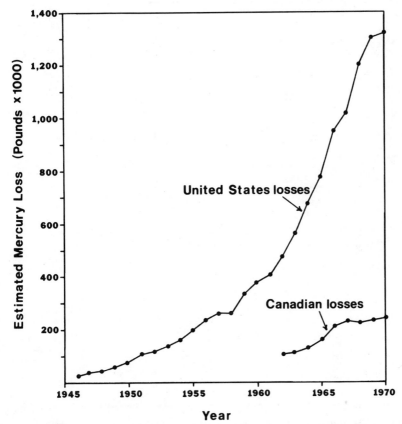

Figure 20. Estimated losses of mercury by the American and Canadian chloralkali industries since 1946 and 1962, respectively.

may actually not have diminished much if some mercury from plants that closed was transferred to others that still operated. Although the consumption rate declined to the 1965 level of approximately 8753 flasks in 1972, American purchases were back up to 15,222 flasks by 1975. However, three Canadian mills converted to diaphragm cells in 1973, and the others reported decreasing mercury losses to the environment. By 1973 the percentage of chlorine manufactured by mercury cells had dropped to 46.8 in Canada (316).

Despite glowing reports of industrial modifications that permitted an initial drop in replacement purchases, mercury losses continued to be

Figure 21. Locations of mercury-cell chloralkali plants in North America.

substantial, although their distribution altered. Canadian replacement purchases dropped from 174,190 lb in 1971 to 93,860 lb in 1972 (177).
If no mercury were obtained from cells that discontinued operation, 100
tons of chlorine were produced with 21.4 lb of mercury in 1972 compared
with 44 lb in 1969. Out of the total discharge, Canadian regulations permitted 0.25 lb to be released into the waterway. Another 3.11 lb were lost
in gaseous emissions, whereas 6.86 lb were retained in solids and sludge.

From mid1972 to mid1973, nine Canadian plants released 20,900 lb
of mercury in solid waste to produce 304,000 tons of chlorine. Since only
0.27 lb remained in the products, more than half of the mercury, 10.9 lb
per 100 tons of chlorine, remained unaccounted for. In other words, half

the 93,860 lb consumed in 1972 disappeared mysteriously. The Canadians suggested that this mercury might have passed through cell-room floors to the underlying soil, but less than 500 lb were found under one floor. With half their mercury waste also unaccounted for, the American chloral-kali producers theorized that it was ventilated to the atmosphere with the hydrogen exhausts (177). Although several plants converted to diaphragm cells, American mercury-cell chloralkali factories continued to purchase substantial quantities of mercury to replace the mysterious losses.

12

NONAGRICULTURAL ORGANOMERCURIAL PRESERVATIVES

PAINT

Despite the greater publicity accorded the hazards of agricultural pesticides and fungicides, they have added much less mercury to the environment than fungicides intended to prevent saprophytic fungi, the surface growths like mold or mildew and internal decomposition such as rot or decay that destroy nonliving materials. Whereas agriculture required 570,000 lb in 1950 but only 106,400 lb in 1971, preservatives that prevent spoilage of paint, fabrics, cosmetics, and pharmaceuticals have been allowed to continue in a number of instances if substitutions would cause a negative impact on the industry. Among the 947,000 lb of mercury used to manufacture pesticides, slimicides, and fungicides in 1968, mildew-proofing paint absorbed about 66%. This amounted to 803,000 lb or 14.4% of the nation's total industrial mercury consumption. By 1972 about 20% of the total went to manufacture organomercurials. Three-fourths of these compounds went into paint preservatives, although they could no longer be added to wood pulp. They had each previously absorbed 3% of the total mercury in fungicides and preservatives.

Although adequate substitutes were not yet available, the EPA suspended the registrations on the relatively small quantities of mercury compounds added to marine antifouling paints in 1972 to prevent leaching more mercury into the waterways (317). Whereas this application had consumed 30,400 lb in 1969, the predicted decline to 24,700 lb by 1974 was actually diminished to 2432 lb in 1973. Thus one minor mercury application begun nearly a century earlier had essentially ended. A mercuric-oxide compound had been applied to protect the bottoms of ships since about 1850, but the organomercurials introduced in World War II greatly increased the durability of paints as well as their resistance to fouling by barnacles and other marine growth (9). After 1940 the American Navy had the hulls of warships painted with mercury antifouling paints. The first was plastic paint Formula Number 142. When a solid mixture of synthetic resins, salts of mercury, lead, and arsenic was melted at about 350°F and kept hot during spraying, exposure to mercurial fumes seemed to be the only hazard.

One team of men suffered the usual symptoms of nervousness, fatigue, anorexia, constipation, and sore gums when they worked several 13-hour days to complete a rush job. Three of the men had hand tremors, and five were hoarse, whereas one worker's face and eyelids were slightly swollen. A silver coating also formed on their gold rings and on the hull of the ship near the portable heating apparatus. However, such risks of overexposure were relatively minor, the effects were slight and reversible, and they could readily be prevented with respirators. Therefore, organomercurial antifouling paints continued to be applied for 30 years. After they were banned, the Navy substituted hot plastic, vinyl, and polyisobutylene antifouling formulations. Some mercury is also used to produce them, but the paints themselves leach out very little. However, these substitute compounds are less effective to inhibit the film of slime in which marine algae grow (318).

Many more tons of mercurials have evaporated or flaked off in paint chips or been discarded in residues of water-based paints. Latexes have become very popular for the interiors and exteriors of millions of buildings because they are easy to apply and clean up, they dry quickly, and they can be produced in a wide variety of colors (319). From 0.01 to 1.5% of the paint may be mercury in concentrations from 100 to 15,000 ppm to prevent the growth of mold. Usually from 1 to 3 lb of fungicide are added to each 100 gallons of paint to insure a mildew-resistant film. In 1969 more than $1 million worth of phenylmercuric salts containing over 727,000 lb of mercury were added to water-based paints. In 1973 mildew-proofing consumed 575,400 lb of mercury. The projected consumption rate was expected to stabilize at about 760,000 lb. In addition, oil-based paints are protected by phenylmercuric acetate, diphenylmercuric dodecenyl succinate, phenylmercuric naphthenate, and phenylmercuric-8-hydroxyquinolate (320).

A ban on these mildew inhibitors would lessen individual exposures to the vapors as well as decreasing mercury losses to the environment. Latex paints initially emit from 0.5 to 25 μg Hg/m^3 of air, but minute quantities of mercury may be discharged for up to 7 years. By then many rooms have been repainted so the supply of mercury is also replaced. Adults are not apt to be affected unless a professional painter with an unusual sensitivity to mercury has such frequent exposure that it necessitates changing occupations. But babies and small children who spend many hours in closed quarters such as freshly painted bedrooms may be exposed to vapor levels that exceed the 16-hour maximum recommended for adults. With the greater susceptibility of children to mercurialism, Donald J. Sibbett has contended that even a limited exposure may cause changes in mood, bursts of temper, and even minor damage to the central nervous system. That could cause tremors of the finger tips, eyelids, or tongue as well as sores in the mouth (321).

The few incidents in which people have been stricken by paint vapors have usually involved some unusual circumstances (322). In a small Kentucky town mercury poisoning was suspected but never proven among 49 people who became ill 3 days after the heat was turned on in the electronics factory where they worked. The boiler was in another part of the plant, and the steam conduits to the coil and assembly room had been painted several months earlier. Since air and urine samples were not taken for several weeks, the results were inconclusive, and mercurialism remained only the probable cause (323).

Misapplications of paint have caused more clearly defined mercury poisoning. When an exterior latex paint containing phenylmercuric propionate was applied to a bedroom and kitchen, a five-year-old boy developed symptoms of mercurialism after sleeping in the bedroom. His condition was first diagnosed as rheumatic fever, but his prolonged irritability promoted a subsequent diagnosis of acrodynia when tests confirmed 90 ppb mercury in his urine (324). The exposure might not all have been from vapors, because the boy could have swallowed or absorbed some mercury through the skin while "helping" his mother paint.

PULP AND PAPER

Mercuric chloride was recommended as a wood preservative in 1790, and organic mercury compounds were adapted in the 1920s to control sap stains, molds, and fungi that destroy lumber. When 16 ethylmercuric and phenylmercuric compounds were tested in 1931, all of them efficiently prevented fungi and other organisms that cause a blue stain or otherwise deface the newly cut wood. Cost, corrosiveness, and danger to human beings or animals were the factors that determined which compound was best applied commercially. They all generated substantial savings. In 1932 dipping lumber into the chemical fungicides prevented degradation from blue stain and molds. They could cost up to $400,000 a year in the southern United States where 6 billion board feet of pine were cut but not kiln dried. Many southern mills subsequently installed dipping vats to protect the lumber from decay when it had to be stored for long periods of time (325).

Workers who treated the lumber were exposed to minor mercurial reactions. In 1940 forty-two men developed skin rashes as they dipped pine boards of varying lengths and widths into a solution containing 6.25% ethylmercuric phosphate. A concentrated stock solution had been prepared by dissolving a 1-lb bag of the powder in 50 gallons of water. After further dilution the dipping trough contained approximately 150 ppm of ethylmercuric phosphate. Although the workmen wore gloves to prevent burns when they worked with the powder, even the highly diluted solution caused a rash within from 3 hours to 3 days. One man's feet also were injured after being wet by the solution. The disability usually lasted from 5 to 30 days. Men who were not as susceptible could work with the solution even without rubber gloves, but the others had to protect themselves with rubber gloves, aprons, boots, and goggles as well as by frequently washing their hands and changing their clothes (326).

Since the hazards were generally minor and the economic benefits great, mercurials proliferated on wood products. During World War II the government specified treatment with phenylmercuric oleate to prevent mold, stain, and decay on wood in defense products. In 1941 an oil-based solution of 2000 ppm of phenylmercuric oleate was shown to preserve the wood with a 100% margin of certainty. Moreover, incorporating the preservative in a light varnish decreased the required number of coats of paint. But skin irritations were common with the 2000-ppm solution, and sensitive individuals were affected by solutions that were half as concentrated. Six-hours exposure could cause a rash, and gloves were no longer recommended, because leaks permitted a continuous irritation. The best preventive was an automatic dipping process, but that was only economically feasible in large operations (327).

The much more hazardous methylmercury compounds were also tested as wood preservatives. A 57-year-old man was poisoned by treating wood with a methylmercury preparation on "repeated occasions" for 5 years. In December 1944 he became numb, dizzy, and had an unsteady gait. As his symptoms intensified, his condition was initially diagnosed as a tumor. Although that hospital had previously treated other victims of alkylmercury poisoning from manufacturing methylmercury hydroxide and treating seed grain, the similarity in symptoms was not observed before the man died (328).

Mercury preservatives were also added to wood pulp to prevent the growth of slime. Moreover, if the caustic soda used to convert the wood chips into raw pulp or the chlorine used to bleach the pulp were manufactured in mercury chloralkali cells, they introduced between 1 and 7 ppm of mercury to be adsorbed on the cellulose fibers (102, 329–330). Thus mercury levels in the pulp might exceed those originally on the wood ten fold or more. Additionally, when salt brine and electrical power were available near the mill, pulp manufacturers occasionally set up small chlorine and caustic soda plants. This could be cheaper than transporting large quantities of those chemicals for long distances. Then the waste from the chloralkali plant would also be dumped into the river, and mercury could be taken in with water added to the pulp at the mill downstream.

Mercury slimicides also preserve the pulp from fungi and bacteria when it has to be stored several months or shipped from one country to another. Wet mechanical pulp may be produced in Scandinavia and Canada to be made into newsprint in Great Britain or the United States. In the storage barrels the high temperatures, humidity, and abundant nutrients in the wood fibers form an ideal substrate in which fungi and bacteria multiply. Drying the pulp before storage could prevent this, but the pulp would have to be slurried again to make newsprint. The cost would increase substantially if both drying and wetting cycles had to be added.

Canadian paper mills added an estimated 3000 lb of mercury to paper manufacturing in 1970. Then 40% remained in the paper, and 60% or about 1800 lb went into the waterways. Some of it was recirculated in new pulp, so approximately 1200 lb of mercury were estimated to flow out in the effluent per year, and it did elevate the levels in fish in some water-ways (121). Three hundred miles north of Montreal, four Indians from the Waswanipi Cree band were treated for mercury poisoning in 1971 after eating mercury-contaminated fish taken from Lake Waswanipi. The Crees had eaten fish with an average of 3 ppm mercury once or twice daily all year, and the four Indians' blood mercury was 10 times the average Canadian's, whereas other Indians in the tribe had 5 times the average mercury concentration. The source of pollution was a pulp mill at Lebel-sur-Quevillon. The mill discharged about 20 lb of mercury daily 25 miles upstream from the waters fished by the Waswanipis (331).

The Canadian wood pulp industry had long been accused of causing the fish to die or taste so bad that fishermen no longer wanted them. When the water downstream from a pulp mill was ⅓ solid matter, largely wood particles, the invisible mercury pollution seemed less threatening. Fines had been levied occasionally during campaigns to clean up the rivers, but when the mills threatened to close, prosecution was usually averted to protect the employees' livelihood. When the mills were pressured not to discharge waste into the river, some of them burned the residues and vented the mercury into the atmosphere.

In the United States mercury slimicides were banned from wood pulp used to make paper products that would come into contact with food in 1965 (149). A year earlier, 163,248 lb of mercury were added to slimi-cides. Since paper manufacturers often did not know in advance which wood pulp would be destined to make paper for wrapping food, many of them subsequently stopped adding mercurial slimicides altogether. There-fore, only 47,044 lb of mercury were consumed for that purpose in 1965. A number of nonmercurial biocides were substituted such as chlorophen-ols, bromoketones, organic sulfur compounds, and chlorine gas.

In July 1969 the Department of Agriculture cancelled the registrations on Algimicin 200[R] and 300[R]. These slimicides had been used to control algae in swimming pools and cooling towers. However, elementary school children in Angwin, California, concentrated an average of 39.6 ppm of mercury in their hair after swimming in pools treated with phenylmercuric acetate in an Algimicin compound for 10 weeks. The pool contained 27 ppm of mercury. Thus people with defective kidneys might concentrate even more mercury if they could not eliminate it normally. The manufac-turer, Great Lakes Biochemical Company, Inc., was allowed to market the products until the ban was reviewed and upheld. On August 7, 1970 it was extended to all mercury compounds in algicides, slimicides, or laundering agents that would be discharged into the waterways (332).

In 1972 the EPA extended the cancellations on rice, laundry, and marine antifouling paint as well as 750 mercurials used in other paints, on wheat and barley crops, on golf courses, and to treat logs and lumber, because they created "a substantial question of safety" (232). However, 600,000 lb of elemental mercury were still added to preservatives each year to mildewproof nonmarine paint and treat fabrics and textiles. These uses were allowed to continue, because their cancellation could adversely affect the products. The same was true where nurserymen and potato growers presumably could lose their crops, because they had no effective substitutes for mercury (333–336).

Many of the remaining mercury pesticides, fungicides, and slimicides were cancelled in October 1973. The EPA contended that any mercury introduced into the aquatic environment could eventually be converted into methylmercury. Moreover, the EPA contended that these products were not needed any longer because effective, economically feasible substitutes were available. The cancellations for 103 registrations ranged from root dips to an antibacterial ceiling tile. Mildew-resistant paint, textiles, and shoe linings were on the list as well as preservatives for cosmetics and topical therapeutics. Of the remaining mercury applications 36 were in paint. Some cancellations went uncontested, and the companies were allowed to dispose of existing stocks (337). Among those contested, phenylmercurial preservatives were allowed in face creams, again because appropriate substitutes were not available to prevent eye infections from the microorganism *Pseudomonas*. These minimal additions of mercury to the environment were inconsequential compared with current and previous industrial discharges.

13

ACCUMULATION
IN THE ENVIRONMENT

Peripheral discharges of mercury and its compounds from manufacturing or burning fossil fuels increase the quantity of the element that circulates through the environment. Thus alone and in combination with other pollutants, mercury is likely to have some impact on the health of all living creatures whether from direct contamination or more gradual concentration in the environment and the food chain.

This assault on the environment began with the burning of fossil fuels for warmth and energy. Citizens protested the excess smoke in nineteenth-century London, and when coal was a widely consumed domestic fuel, household members may have been exposed to up to 1 g of volatilized mercury per ton as the fumes spewed out over the housetops. Although one study reported that mercury levels in human tissues preserved between 1913 and 1970 have declined in recent years, most evidence indicates that the opposite is true (338). What is clear is that neither the early antismoke laws nor the advent of technical innovations like the diesel engine could stem the exhausts from industrial waste and fossil fuels as the layers of atmospheric pollution thickened over the cities (339).

The industrial revolution and the expanding world population both called for more fossil fuels, first coal and then oil. After the steam engine was invented in 1705, iron was smelted in coke-fired furnaces. James Watt improved the steam engine in 1765, and 20 years later this power source was propelling equipment in English cotton mills. The increasing consumption of coal steadily poured more mercury and other waste into the atmosphere to settle over the land and water. Whereas between 1660 and 1700 England generally mined only from 2.25 to 2.5 million tons of coal per year, the tonnage rose to 6 million by 1700 and to 10 million a century later. The demand for coal subsequently increased from 20 to 100 fold in the nineteenth and twentieth centuries.

The American colonies slowly industrialized, and the United States consumed only 15,000 tons of coal in 1820. Then by 1850 the demand increased rapidly to 7 million tons a year, 33 million tons by 1870, and 357 million tons by 1926. Coal consumption had exceeded 559 million tons per year by 1950, although oil and gasoline replaced it in many instances, particularly in the expanding transportation network. Although steam locomotives became operational about 1784, the beginning of the railroad era is usually dated from 1830 when 200 charters were granted to expand the system.

Even more than the cotton gin, another of Eli Whitney's inventions, interchangeable moving parts, accelerated manufacturing during and after the American Civil War. Then, in 1879 the industrial revolution was accelerated by the addition of electric lights and the dynamo, still usually fueled by coal. The diesel engine created a new demand for petroleum products in the 1890s, and the first automobiles began emitting exhaust fumes in 1900. Thus at the turn of the century all the basic equipment for transportation and heavy industry was in operation and became more sophisticated to meet greater demands for manufactured products as the world's population expanded from 900 million to 3.7 billion people between 1800 and 1971. The American population increased 42 times during this period and consumed a much greater proportion of the fuel and raw materials than less industrialized countries while absorbing the waste into the environment.

Although burning coal was recognized as a significant source of mercury in air and rainwater by 1954, more of this readily available fossil fuel continued to be consumed than ever before (340). In 1968 American industries released almost 6 million lb of mercury to the atmosphere, 10 times the Canadian emission, for a combined total of approximately 6.9 million lb per year. In 1969 the world released approximately 23.4 million lb of which 22 million lb came from industry. Estimates of the losses to the atmosphere by burning fossil fuels such as coal, oil, and natural gas range up to 1.4 million lb per year (341).

The quantity of mercury in combustion gases depends on the levels in the fossil fuel. Samples of coal from American mines ranged from 0.070 to 33 ppm (342–343). They are greater in crude oil and natural gas from deposits located near cinnabar ore. Some of the highest values ever reported for mercury in crude oils, from 19 to 21 ppm, were from California's Cymric Oil Field where the brine also retained from 0.020 to 0.20 ppm of mercury. Although these individual proportions are small, natural gas and oil from the Cymric Oil Field alone emit enough mercury to fill thousands of flasks (344).

Mercury exhausts are only a small portion of the complex mixtures of fumes and particles that range from hazardous to relatively innocuous compounds in the increasing layers of smog that spread from the cities into the countryside. After the first major American smog alert caused 6000 illnesses and 20 deaths among the 14,000 citizens of Donora, Pennsylvania, in 1948, the incident prompted better enforcement of smoke-control laws. By then, however, they were grossly inadequate to monitor the air pollution that had enveloped all major cities. In addition to the human toll, in European cities intricate carvings that remained intact on ancient edifices for centuries have begun to blur as their surfaces are gradually defaced by the acidic haze. For instance, four bronze horses stood atop the center door of St. Mark's Cathedral in Venice for centuries after their arrival from Constantinople in 1294, but they had to be moved indoors in the 1960s to prevent further deterioration.

Contaminants have spread so widely through the environment that attempts to determine how much human activities have raised natural background levels of mercury have to be viewed with some skepticism. In 1970 mercury concentrations in rock and soil samples collected at 50-mile intervals across the United States averaged 71 ppb compared with 50 ppb for the earth's surface as a whole (345). The levels were generally higher east of the ninety-seventh meridian and averaged 96 ppb compared with 55 ppb to the west. Summit County, Utah, tested highest at 4600 ppb, whereas several states had areas with over 1000 ppb in the soil. Even where samples were taken 8 in. beneath the surface of the ground, the natural background levels of mercury were exceeded in cultivated soil, probably because it attracts more atmospheric fallout and often has received agricultural fungicides as well.

In addition to industrial exhausts from manufacturing processes and the combustion of fossil fuels, natural releases by weathering may add between $\frac{1}{2}$ and 11 million lb of mercury to the atmosphere each year (342, 346–

348). In a continuing cycle the element leaches out of the ground into the waterways and may vaporize from the soil and water into the atmosphere. Then it precipitates and is washed out in rainwater to settle again in the waterways or upper strata of the soil where it may bind permanently or vaporize again. The output from industry is added to this ongoing natural cycle of vaporization and condensation.

As with the concentrations in soil, little attention was paid to mercury levels in water or air before 1970. Early test results for mercury in ocean waters have to be questioned, because the sulfuric acid used as a reagent undoubtedly contained mercury impurities (349–350). However, in 1877 the presence of mercury was confirmed in French mineral waters (351–353). Despite contamination the levels of mercury generally remain quite low in watercourses, because the element readily adsorbs on particles and settles to the bottom.

After a fish kill at the Boone Reservoir in 1968, the drinking-water supply for five Tennessee Communities was threatened with a toxic mercury compound. Then the Public Health Service set a maximum of 5 ppb as the acceptable level of mercury in drinking water and tested supplies throughout the United States in 1970 (354). Most of them were under 1.0 ppb. Ten measured from 1.0 to 5.0 ppb, but only two exceeded the 5.0-ppb limit for potable drinking-water supplies. More mercury was measured near industrial outlets, but the proportion quickly declined a short distance downstream. Even where no industrial sources were evident and the levels were low, they still varied substantially, apparently because of natural background leaching and airborne fallout from nonpoint industrial discharges as well as because of differences in acidity, temperature, redox potential, natural chelating agents, and the occurrence of complex ions.

Since drinking water is also drawn from underground aquifers, some of the mercury added to the soil from airborne, agricultural, and land disposal discharges may eventually percolate into the subsurface waters, although most of it binds with the soil, evaporates, or leaches into surface streams and rivers that transport it to the oceans.

When tuna and swordfish tested over the 0.5-ppm FDA standard in 1970, the possibility loomed that the oceans could also be accumulating heavy metals to levels that could be toxic for some organisms. However, when samples taken from the North Sea in 1934 were compared with 1971 analyses of the Northeast Atlantic and the English Channel, it appeared that the enormous size of the oceans at least insured that the contaminants accumulate so gradually that destruction is a longtime proposition. The 1934 samples were 0.03 ppb, whereas later they ranged between 0.14 and 0.021 ppb (285,355). Although the oceans are now estimated to contain from 100 to 300 billion lb of mercury, samples of sea water usually range from 0.03 to 0.3 ppb depending on the area and depth at which they are taken. If the oceans receive all the mercury pollution, at the present rate it has been estimated that their total concentration will double in 125 centuries.

Large deposits of mercury already rest in the sediments of relatively limited areas near the outfalls from chloralkali plants, pulp and paper mills, and other discharge sites. Downstream from sewage treatment plants, the levels rapidly dropped from 6800 to 50 ppb (9). Where municipal sewage wastes were dumped into the ocean, the nearby sediments were 2 times richer in mercury than those farther from the outfall. An estimated 58 tons of mercury rest in the top 12 in. of mud and silt in San Francisco Bay where the mercury content ranges from 20 to 2000 ppb with a high value of 6430 ppb (356).

Although the mercury concentrations are mounting because of municipal and industrial contamination, they are generally nowhere near the 2,010,000 ppb (wet weight) measured in the sediments of Minamata Bay. This small watercourse contains an estimated 1.32 million lb of mercury (357). In some cases, perhaps because the sediments have been washed into the ocean, the mercury levels remain relatively low despite prolonged contamination. London clay from the Thames estuary had essentially the same 10 ppb of mercury in fresh and 50 million-year-old specimens (358). In contrast, samples of sediments from waterways in the Western Hemisphere more often seemed to reflect the increasing mercury pollution. In cored specimens from Lake Ontario, Canada, the background of approximately 358 ppb gradually increased between 1900 and 1940 and then quadrupled to 1400 ppb by 1950. After that the mercury levels fluctuated, and the trend reversed downward to approximately 1300 ppb in 1968. Near-shore sediments averaged 355 ppb, whereas the total lake basin averaged 997 ppb. On the industrialized southern shore of the lake and in the western Niagara basin, the sediments concentrated approximately 2100 ppb of mercury. Unless natural sedimentation or artificial coverings confine the element, it can be methylated through natural processes in the waterways to be released and concentrate in the fish over decades or even centuries.

III

MERCURY IN MEDICINE

14

THE HISTORY OF MERCURY IN PHARMACEUTICALS

Although the ancient Greeks generally regarded mercury as too poisonous for internal consumption, Aristotle recommended the element, diluted with saliva, to treat certain skin conditions. Then Pliny the Elder incorporated mercury in an ointment in the first century A.D., and other physicians added mercuric chloride and other mercurial compounds to ointments or salves applied to treat a number of skin diseases. Dioscorides Pedanus also favored cinnabaris to cure eye diseases, staunch blood, and heal burns. Although Paulus of Aegina recommended mercury for certain intestinal troubles, the element's heavy weight was long considered a disadvantage for medicines taken internally, because it was erroneously thought to pass through the body in a lump. Therefore, large quantities of less dense fluids such as milk or wine with wormwood were usually also prescribed to dilute the mercury (359–360).

In the Far East Indian physicians may have included mercury in medicines as early as Buddha's time, about 500 B.C.; this certainly was true by or shortly after the birth of Christ. The ancient Hindus believed quicksilver was an aphrodisiac, and they consumed large quantities in the vain hope of attaining renewed sexual vitality. Not surprisingly, many accounts describe such symptoms of mercurialism as loose teeth, salivation, and disintegrating gums. Indian medical secrets involving mercury were carried to Persia, and Arabic physicians assimilated both Indo-Persian and

177

Greek practices after the Persian empire was conquered in 650 A.D. The Arabs, in turn, devised new techniques of mercurial therapy, particularly complicated ointments. Mesue the Older (died 857 A.D.) used one to treat body lice, scabies, itch, impetigo, pustules, and leprosy.

The Europeans inherited many misconceptions about the effects of mercury on the human body as well as some formulas for ointments from the Arabic physicians, and some of each continued into the twentieth century. In Europe research was hampered by the powerful influence of the Roman Catholic Church, although some Greek and Roman medical theories were adopted. Without this religious deterrent, Arabic scientists continued to outstrip the European study of medicine until the two cultures met during the Crusades and the Moorish wars in Spain. Then European scholars, among whom Constantinus Africanus was one of the best known, translated large numbers of Arabic medical works into Latin. By 1180 A.D. Italian physicians had adopted Arabic prescriptions for mercurial ointments with which they treated vermin, scabies, and body lice. The European innovation was to heat the skin after applying the mercury ointment to increase the effect of the medication, not to mention the likelihood of mercurial intoxication. Theodoric of Cervia (1205–1298 A.D.) described four early modifications of these Medieval heat and mercury tortures. With all of them, the patient underwent a 6-day preparatory treatment of laxatives and baths to increase his receptivity to the ointment and heat.

Both Theodoric of Cervia and Guy de Chauliac observed that mercury intoxication caused excessive salivation, a relationship their Arabic predecessors either had not understood or at least did not encourage. But European physicians concluded that some salivation was desirable to show that the mercury treatment was taking effect. However, excessive salivation was initially regarded as an alarming symptom that the treatment might be harmful. Consequently, excessive salivation was treated with ordinary honey or rose honey, and if that did not help, the mercury treatment was discontinued.

Then salivation was also turned into a medical technique, and large quantities of saliva were induced to remove excess phlegm that was thought to accumulate in the body because of diseases such as gout. The notorious "inunction cures" ultimately took the health and sometimes the lives of thousands of Europeans. They remained popular throughout Europe even into the twentieth century, perhaps because of the continuing influence of Medieval physicians like Phillipus Theophrastus Aureolus Bombastus von Hohenheim (1493–1541 A.D.), more commonly known as Paracelsus. Although he favored the inunction cures, Paracelsus was far ahead of his colleagues in his belief that life is essentially a chemical process so changes within the body should be treated with chemicals. He shared the contemporary belief that the body is composed of mercury, salt, and sulfur, but he was the first Medieval physician to attempt a comprehensive explanation of the origins of disease and the effects of mercury on the human body.

Paracelsus espoused the Medieval doctrine of correspondences wherein mercury is associated with the moon and cold in contrast to man's nature which is warm. He theorized that every disease is a frost that leads to death, and mercury applied to the skin can create "winter" in which the victim shivers, his teeth chatter, and the heat retreats to the interior of the body. When it can flee no farther, the intense heat may cause death through suffocation. As the fire consumes the bones, marrow, blood, flesh, and other parts of the body, putrification destroys the lungs, liver, stomach, brain, or whatever the fire acts upon. In the process secondary diseases may also arise to attack specific parts of the body and cause toothaches, madness, or paralysis. Because cold causes all diseases and can induce purgation, putrification, and chilling, Paracelsus contended that mercury compounds should be taken internally to drive the cold out. He believed mercury poisoning was caused if the metal was deposited in the body in a stable form. Then the heavy metal could gravitate into one place, a hip joint, for example, where it had to be forced out by applying a corrosive plaster that would produce an ulcer in 2 or 3 weeks and revive the dead mercury. After it left the body, baths with either herbs or sulfur were prescribed to complete the treatment (361).

Despite the hazards, Paracelsus and other mercurialists contended that inunction treatments were highly beneficial. They devised new applications for cinnabar fumes, plasters, corrosive sublimate washes, pills, and other mercurial preparations. They also thought the body's tolerance for mercury was increased by laxatives, baths, starvation, and intense perspiration. Despite the physician's best efforts, the patient's deteriorating condition also made him vulnerable to other illnesses. The inunction and salivation cures caused death indirectly.

A growing cult of antimercurialists opposed having mercury applied or ingested when it would cause diarrhea, loss of appetite, and thirst or more serious problems. By the sixteenth century mercurials were also known to affect the kidneys adversely as well as to cause paralysis, vertigo, loss of consciousness, twitching, epileptic seizures, and disturbances of speech, vision, hearing, and mental powers. Nonetheless, new medical applications for mercury have continued to be introduced, whereas others have passed out of fashion.

15

OBSOLETE OR DECLINING MEDICAL TREATMENTS

SYPHILIS

In 1972 the American public was shocked to learn that treatment for syphilis had been withheld from a group of black men in Alabama over the previous 40 years as scientists observed the ravages of this ancient disease on the human body. Many of the subjects in the Tuskegee study were not even told why the nurse periodically collected samples of their blood (362). The experiment seemed pointless after World War II when penicillin was discovered, since it cures the disease in its early stages. However, when the experiment was instituted in the 1930s, the symptoms of the disease and reactions to treatments with mercury and arsensic were still unclear. That the three stages of syphilis were all part of one disease had been confirmed long before, and in the nineteenth century Kussmaul had demonstrated that chronic mercury poisoning and late syphilis were separate entities. Even this scant information had been obtained at the cost of additional suffering because of misinformation and maltreatment of thousands, even hundreds of thousands of people over the centuries.

Mercury was used to treat syphilis long before medical students begot the rather brutal jest that "a moment with Venus may mean a lifetime with Mercury" (363). The origin of the disease is unknown, although similar symptoms were reported in ancient Egypt where the stages of the disease were not identified as parts of a single ailment. The earliest reports of a disease like syphilis came from China where the Emperor Huang Ti (circa 2657 B.C.) was rumored to have tried to cure this affliction with mercury (364). Nor can diseases described by early physicians like Pliny the Elder and Avicenna be confirmed as syphilis. However, later European doctors treated syphilis with some of Avicenna's mercury ointments.

The first documented epidemic of the "French disease" occurred shortly after Columbus returned to Spain from America, and mercury was a primary medication as this terror swept Europe. Initially, the coincidence of timing with the return from the New World was not noted, and Columbus' sailors were not blamed for the new plague until later. Early accounts of the disease and its cures were sung in poems more closely related to myth and legend than the pursuit of health or sex.

Francisco Lopez de Villalobos was still in school when the wave of syphilis hit Spain in 1493. In a long poem he contended that God originally manifested this plague to punish Egyptians for their venial sins. The symptoms that de Villalobos described now seem more common to the cure than the illness. The leaden complexion, erratic sleep, discolored lips, and dragging feet have long been identified with mercury poisoning, but de Villalobos' description muddied the distinction between the disease and its treatment for 4 centuries.

The plague appears first upon those parts restricted
They are free from pain and hard as hand that callused.
Eyes black rimmed, leaden face and head with pain constricted
With weighted shoulders the victim stands convicted
To strange short dreams and sleep that is forever lost.

Black lips and heavy lids ill favor the visage
Feet do drag unwillingly their duties to perform.
Lack lustre eyes, through hazy fog unseen gaze
And he performs in duty but a tithe of other days.
All these signs with Egypt's plague do well conform (364).

At the beginning of the sixteenth century, Italian physicians were treating syphilis with mercurial ointments that included everything from friction to frogs. In 1500 A.D. Torrella condemned applying mercury ointments with friction, because he believed the rubbing and excess salivation would not rid the body of evil fluids, and he cited as evidence that this treatment had already killed two cardinals and Alfonso Borgia. One of Borgia's doctors, the Spanish physician Petrus Pinctor, also believed the Cardinal of Segovia had died of a mercury overdose, but he asserted that mercury ointments soothed the pain and reduced the number of pustules in a short time. Thus Pinctor favored treatment with 1½ oz of quicksilver slaked in the patient's saliva, and he optimistically concluded that this element was the most valuable cure for syphilis when applied in correct, moderate doses.

Although mercurial fumigations were becoming fashionable in Italy in 1504, the Genoese physician Jacobus Cataneous also recommended friction treatment after the syphilitic patient was washed with solutions of mercury sublimate. Additional medication included diet, bleeding, purging, and applications of vipers' flesh (363). In 1514 Giovanni di Vigo contended that mercury friction and plasters should be continued until the patient's teeth ached. This signal to stop the cure usually occurred after a week of applications. Although di Vigo did not encourage his patients to expect permanent cures, his prescription for mercurial plasters remained popular, and his poultice, supplemented with live frogs and earthworms, was still being recommended in some medical textbooks during the nineteenth century.

Poet-physician Hieronymus Fracastorius reviewed alternative theories of the origin of syphilis as well as treatments in his Latin poem "De Morbus Gallico" composed in hexameters in 1530. He speculated that the disease might have formed in the Earth's atmosphere through astrological means at the conjunction of the planets Jupiter, Saturn, and Mars. Then the division of Cancer corrupted the air breathed by mortals who developed syphilis. Or the gods Diana and Apollo might have inflicted the disease on Ilceus for killing a sacred stag. Ilceus is treated by bathing 3 times in an underworld lake that flows with liquid silver. Or the disease might have originated in the New World, and Fracastorius describes guaiacum, the Holy Wood cure from Haiti. He calls the disease Sypholus after the goddess of Ammerce who revealed the secret of guaiacum (365). A similar story of how the French disease came to be called "syphilis" contends that Syphilis was a shepard stricken with this illness because of irreverence toward the sun. He also was supposedly cured with Holy Wood (236).

Whatever its origin, syphilis was not the name that commonly identified this disease for another 3 centuries (365).

Although the gods might be cured in a stream of liquid silver, for ordinary citizens Fracastorius recommended traditional ointments composed of arsenic and mercury. The poem includes complicated prescriptions for preparing mercurial salves to be applied after purgation and bleeding. The mercury would then be rubbed into the skin with friction over 10 days in treatments so drastic that "you will see the odure of the plague flow from the mouth consciously in the noisome spittle, and marvel at the abundant stream of diseased matter that falls at your feet" (363). Fumigation was an even more drastic tactic, and Fracastorius describes how the patient was placed on a chamber pot with his entire head and body covered with blankets while cinnabar was burned beneath them. As he breathed the vaporized mercury, the patient would perspire freely, and continued treatment could cause his teeth to loosen and his bones to deteriorate. These treatments are described in a section of the poem aptly titled "The Sinister Shepard" (364). Later, ovens were devised in which to steam the victims until perspiration poured fluids from their body to augment the salivation (see Figure 22).

In 1539 forty-six years after Columbus' ships returned to Palos Bay, Ruy Diaz de Isla blamed the plague on the sailors. In his poetic description of the disease and its treatment, *Tratado Contra el Mal Serpentin*, de Isla noted the three stages of the disease and stressed that the time to cure it was before the outbreak of such later symptoms as the buttonlike skin lesions. He did not think the second stage was contagious, although he knew the third was usually fatal. De Isla prescribed the application of ½ oz of mercury per day for 9 consecutive days to be followed by silver and then cow fat if the skin ulcers appeared anyway. For sores in the mouth, he advised gargling with mercury and rinsing with goats' milk. In the second stage of the treatment de Isla recommended no physics and no exposure to drafts for at least 30 days while the patient was rubbed with mercury ointment to induce salivation. This treatment could last from 5 to 9 days. Altogether, 15 oz of mercury were rubbed into the patient's skin in 18 applications. Although the disease was usually fatal if it reached the third stage, de Isla recommended continuing the mercury ointment to arrest the symptoms at least temporarily. At this point he recommended that the patient abstain from sexual intercourse to conserve energy (364).

Another poem written by Jacques de Bethencourt in the first half of the

Figure 22. In the sixteenth century treatment for syphilis required that the patient sit in a barrel containing mercuric sulfide (cinnabar) that was heated to produce vapors. From *Folke Henschen, Sjukomarnas Historia,* 1962.

sixteenth century personified the two rival syphilis cures. Mercury and Guaiacum argue at length about their relative value, but the controversy is finally resolved in favor of the older protagonist, Mercury. By this time mercury medications were also being prescribed to be taken internally. Pierre Mattholi recommended one in 1533, and Paracelsus had a special preparation, "turpethum minerales," in which mercury was digested with oil of vitriol (sulfuric acid) and distilled with alcoholic spirits. This remedy was supposed to attack the disease toxin directly instead of by causing salivation and diarrhea.

By 1545 many syphilologists had concluded that guaiacum was not the New World's answer to syphilis, but some physicians like Ulrich von Hutten continued to favor Holy Wood cures over mercury. He contended that if the disease came from the New World, the antidote should also. Von Hutten grimly noted that scarcely one patient in a hundred survived the treatment after being massaged with mercury and roasted in an oven or smothered in bed to release perspiration. If the patient were kept in bed, other members of the household assisted in the treatment by lying on top of him for hours to generate the necessary warmth and perspiration. Relapses were constant and nearly always fatal (363).

Benvenuto Cellini's (1500–1571 A.D.) autobiography also recounts a dramatic failure to cure syphilis when a famous Florentine physician, Giacomo Berengario Carpi, bilked desperate patients out of exorbitant fees after fumigating them with mercury vapors until the disease seemed to clear up. It recurred after he was paid and left for Rome. Otherwise, Cellini observed that Carpi's former patients certainly would have murdered him. Cellini himself was one of the few who recognized the symptoms of mercury poisoning when he was taken violently ill after eating a small portion of a salad while dining with a farm family outside Florence. Apparently, the farm wife had added corrosive sublimate (mercuric chloride) to the salad dressing, because Cellini's physician confirmed the acute mercurial poisoning from its effect on his gastrointestinal tract.

By 1869 corrosive sublimate was also being administered as a cure for syphilis. In malignant cases small doses (about 8 mg) were injected as often as twice a day, and a course of treatments often consisted of from 10 to 20 injections. This practice caused such great pain that it had to be discontinued, and corrosive sublimate later had a more proper if painful role as a favored suicide weapon. Soluble salts of mercury have also been prescribed more recently to be injected or consumed despite the abscesses and severe salivation that they often caused (364).

As medical science progressed, syphilis sufferers were sometimes placed on a high protein diet and their bodies were cleansed of sores and fumigated in a tub containing mercuric sulfide (366). Chronic mercury intoxication remained an occupational hazard for physicians who treated syphilitics by rubbing mercurial unctions into their skin (367). Despite the new treatments, syphilis was so widespread in the seventeenth century that prostitutes were occasionally sent out as a military weapon to "lay waste" to invading armies. Some authorities have speculated that the ravages of the disease and its treatment with mercurials, both of which destroy brain cells, could have undermined the mental capacity and emotional stability of the entire population of the Western World. No comparative studies of intelligence before and after the epidemics are possible, but the widespread exposure can be amply documented, because people openly admitted their affliction and exchanged prescriptions to cure the French disease which was considered more of a nuisance than a disgrace until well into the eighteenth century.

Some of the behavioral abberations such as the well-known exploits of the Marquis de Sade as well as some European kings and lesser nobility may well have reflected the mental deterioration caused by the advanced stages of syphilis and its treatment with mercury. Since the third stage of the disease causes the body and brain to deteriorate, even more mercury was freely prescribed as a last hope before an otherwise certain doom. Consequently, since Napoleon was in the third stage of the disease when he was incarcerated on the Isle of Elba, his death by poisoning may have been an accidental side effect of mercury and arsenic treatments rather than intentional murder.

The obvious social disadvantages of syphilis caused more people to try to hide their contagion in the eighteenth century, and the mercury cures also had to be altered. In 1773 Baron Sean-Charles Le Febure proposed to the French government that he would guarantee to cure all "venereal subjects of the kingdom or pay a forfeit of 24 livres a head." The remedy that he guaranteed to be infallible was a composition of chocolate and corrosive sublimate. The primary advantage of this cure was that it could be taken in the presence of others without arousing their suspicion.

A husband may take his chocolate in the presence of his wife and she will never suspect the secret. She can even take it herself without the slightest idea that she is taking an anti-venereal specific. By these innocent means peace and concord will be established in the family circle. A father may take it in the bosom of his family, a son or a daughter may do so in the presence of their parents and even if it should be taken by those who have no need of it, it will do them no harm, and they will not find it distasteful. A traveller can carry his chocolate about with him and need not load himself with bottles and phials, which always attract undesirable attention (363).

The French government rejected the Baron's proposal, which was probably more favorably received by the public at large as secrecy became the hallmark of genteel refinement. In the early nineteenth century syphilis was even reclassified as a misnomer for a collection of other diseases that were all aggravated by mercury treatments. When the disease was reinstated as a separate entity, new mercury treatments reflected advances in theories about human health.

The advent of the germ theory prompted the idea that mercury acted as a tonic that restored the human system to its normal vigor and poisoned the germ indirectly (368). The element was also thought to cure exactly what it could cause: lost motor control and paralysis. Mercury treatments were also dubiously justified by the belief that they had to make people better instead of worse or they would have succumbed, and many did. This was still true when the colloidal mercury and inorganic mercurial salts employed in syphilotherapy since 1493 were replaced by organomercurials in the latter half of the nineteenth century (369). But the side effects changed. In 1887 the syphilis remained as new poisoning symptoms were induced after two injections of diethylmercury. This prescription was discontinued when animal experiments established that the alkylmercurial compound was highly toxic (370). Nonetheless, at the beginning of the twentieth century, mercury treatments for syphilis were held in as high esteem as ever.

Mercury has a high reputation and is as indispensable in the cure for syphilis today as it was four centuries ago. It has as yet no substitutes. We appreciate every day, more and more, how thoroughly it can be depended on to do the work we ask of it (370).

Since mercurials were common in European medication in the fifteenth century even before the first syphilis epidemic, John H. Stokes, Chief of Dermatology and Syphilology at the Mayo Clinic in Rochester, Minnesota, contended in 1920 that they had been proven effective if the doses were regulated accurately, something modern physicians had learned to do. "No matter in what form it is used, the action of mercury on syphilis is one of the marvels of medicine," according to Stokes. He contended that it could clear up the most terrific eruption with scarcely a scar and transform a bed-ridden patient into a seemingly healthy man or woman, able to work, in the course of a few weeks or months. Mercury treatments were available in pills, ointments, or injections of oil-based formulations which Stokes considered the most convenient and private way to cure the disease. Some doctors recommended daily injections, whereas others thought treatments once or twice a week were sufficient. The major disadvantage was believed to be the discomfort that followed each injection for a few hours. Injections were preferred over pills or liquids taken orally because the patients remained under the doctors' close observation. Physicians feared that small oral doses of mercury might hide the evidence of syphilis without curing the disease. Even if the doses were as large as the stomach could tolerate, Stokes contended that "it seems almost impossible to give enough mercury by mouth to effect a cure." Patent medicines also included the element, but these do-it-yourself syphilis cures were held in the lowest esteem by the legitimate medical profession. Ointments were still preferred as the most reliable means to achieve a permanent cure, but they had many disadvantages.

When the fluid metallic mercury was mixed with a fatty substance to form a salve, the residues on the victim's skin and clothing were difficult to keep secret. Moreover, a great deal of time was required to rub mercury into the skin from 4 to 6 times per week. This amounted to several hundred rubs over 2 to 3 years (370). Since much of the treatment's potency was still believed to stem from inhaling the vapors driven off by the patient's body warmth, overdoses of mercury still caused teeth to fall out and bones to rot. Even in the 1930s syphilologists frequently induced inflamation in the mouths of patients who were sensitive to mercury. Since poor dental health increased this irritation, a trip to the dentist was generally recommended before treatment for syphilis began (371).

As blood carried the mercury to all parts of the body, the poison was supposed to kill germs, stimulate the production of antibodies, and build the body's resistance. Since the mercury was thought to act slowly, it seemed natural that syphilitic sores continued to appear, especially in the mouth, throat, and around the genitals even after treatment was underway. And physicians recommended that treatment be continued long after the obvious signs were eliminated in case all the germs had not been killed. They thought the major disadvantage in the wonder drug was that the symptoms would vanish so the patient would think he was cured and discontinue treatment. Physicians feared that too little mercury was more often given than too much. "Mercury therefore carries its disadvantages with its advantages, and by its marvelous but transient effect only too often gives the patient a false idea of his progress toward cure" (370).

Since relapses did occur despite mercury's reputation as a cure-all for syphilis, there were new innovations. Neosalvarsan[R] was introduced in 1914, and the French reintroduced bismuth as a superior medication in 1921. But the new drugs seemed less effective than mercury, the longer they were used (372). Therefore, new mercury ointments were also devised. Louis A. Duhring recommended that 1.28 to 1.92 g of ammoniated mercury in an ounce of ointment be rubbed into the skin. The popular blue ointment, oleate of mercury, became a standard treatment for syphilis all over the world (373). In 1927 a mercury derivative dissolved in a compound of chaulmoogra oil was touted as a two-in-one "wonder drug" to cure both syphilis and leprosy (374).

Despite such far-fetched claims, some forms of syphilis did seem to be slowed by mercury, bismuth, arsenic, and their iodides. Organomercurial injections caused tertiary syphilitic skin lesions to disappear. Mercury salts were thought to help when injected in the muscles despite the painful side reactions, and since the benefits seemed to be enhanced in victims with high fevers, the old heat treatments were resumed. William Wallace's "syphilitic cocktail" also became very popular. This exotic mixture of mercuric iodide, potassium iodide, and syrup of sarsaparilla was considered an excellent remedy for the third stage of syphilis and long placed a fair second to penicillin as a favored treatment (364). At the beginning of World War II, penicillin was proven effective against the early stages of syphilis, and medical experts eventually decided all the mercurial "cures" had little effect except in the late stage of syphilis when they might help put the patient out of his misery more quickly. All for nought were the charity clinics of Europe and England filled with patients with great stinking ulcers that exposed the bone, noseless faces with powerful, repugnant breaths, and toothless mouths with sore and bleeding gums.

CONTRACEPTIVES

Although a dismal failure as a cure for syphilis, mercurial compounds have been more effective in contraceptives and douches to prevent or cure vaginal infections and to induce abortions. In ancient Greece women swallowed large quantities of elemental mercury, apparently without ill effect, to cause abortion. Later they used other mercurial compounds to end unwanted pregnancies or commit suicide. A higher incidence of natural abortions has also been reported after treatment with mercurial medications. Consequently, in the eighteenth century women were given only palliative treatments for syphilis until their babies were born (363). Whether the mercury or the disease caused the abortion was long debated, but it could be induced by mercury alone.

A 31-year-old, divorced mother of eight children demonstrated this in 1958. She swallowed five tablets (about 0.5 g each) of mercuric chloride and developed such symptoms of acute inorganic mercury poisoning as nausea, vomiting, abdominal cramps, diarrhea, and suppressed urination. The woman aborted her child 13 days later, culminating a pregnancy that had previously appeared normal. Admitted to the hospital in her tenth week of pregnancy, the uterus had decreased to the size of a 6-weeks gestation when the child aborted. Mercury could still not be conclusively demonstrated to have caused the abortion despite high levels in the fetus. However, when mercuric chloride was injected into the stomach or muscles of pregnant rats, the unborn litters either aborted or died.

Both organic and inorganic mercury compounds are spermicides that have often been included in contraceptive tablets, jellies, and vaginal douches. After a London gentleman named Condom invented the prophylactic known as "English frockcoats" to prevent the French disease, one enterprising libertine prepared a pomade of malleable mercury and saliva "to anoint his genitalis before approaching the accomplices of his debauchery . . ." (363). And vaginal douches were devised as contraceptives to control infections of the female reproductive organs, some with disastrous results. A 27-year-old woman died in 1916 after she inserted a tablet containing 3.75 g of mercuric chloride directly into the vagina instead of dissolving it in the recommended quantity of water (373). However, in 1933 when a very dilute solution of phenylmercuric nitrate was applied to the cervix and vagina, in tests on 70 women all but one vaginal infection improved rapidly, even among the 13 with acute gonorrhea. Proper proportions of the solution not only did not irritate the vaginal mucous membrane, but it was colorless, odorless, and did not stain. Bonus qualities were that the solution did not deteriorate when stored, nor did it corrode surgical instruments (375). Even after birth control pills were available, several mercurial preparations remained on the market because some women could tolerate them better than the pills.

DIURETICS

Whether from tensions caused by the modern society or an excess of cholesterol, middle-aged American males seem particularly susceptible to cardiac failure. Although mercury is not used to treat this condition, over the past 70 years small quantities have been widely prescribed in diuretic compounds to relieve the accompanying congestion of swollen arms, ankles, and other parts of the body when excess fluid is retained in the tissues. Nonmercurial compounds are now more often recommended to avoid sensitivity reactions to mercurial diuretics, but they are still prescribed when the congestion is not controlled by rest, digitalization, salt restriction, and the nonmercurial diuretics.

The action of mercurial diuretics on the human body is still not completely understood, although Paracelsus described calomel's diuretic effect in the fifteenth century. In 1799 this was observed to increase when calomel was administered with digitalis. After 1886 small doses of calomel purgatives were regularly administered to heart patients in hopes that a diuresis could be induced and mercury poisoning avoided. The side effects ranged from mild to disastrous (376). Milder reactions were skin irritation, fever, anemia, and weight loss. In the 1890s if the diuretic initiated an irregular jerking pulse, it was thought that metallic mercury globules backed up the current of blood until they were forcibly overcome. The most extreme reaction was paralysis and death (377). Since the diuresis was often inadequate and the side effects were erratic and dangerous, calomel diuretics were discontinued after trials at numerous clinics.

More effective organomercurial diuretics became available after 1917 when the Bayer Company originally marketed Novasurol[R] to treat syphilis. After an Austrian army doctor supplied samples to a clinic in Vienna, the diuretic effect was observed, and the compound was tested on cardiac patients (378–379). After Novasurol[R] successfully reduced their excess fluid, a second organomercurial diuretic, Salyrgan[R], was marketed in 1924, followed by several others. Since the effect but not the cause was understood, the appropriate dosage of the different compounds as well as the best sites of administration had to be determined by trial and error. Individual reactions depended on the patient's physical condition, sensitivity to mercury, size, and age. Nonetheless, between the first fatality directly attributed to Novasural[R] in 1925 and 1942 when a list was compiled from the medical literature, only 26 deaths had been reported from organomercurial diuretics. Altogether, they have caused less than 100 known fatalities, although they have been administered to thousands of people. In many cases other illnesses like cardiac failure, kidney damage, or competing medicines such as digitalis were contributing causes. Sometimes the patient's condition was so poor that physicians believed the administration of any drug could trigger a fatal reaction. To prevent negative side effects, however, new compounds were occasionally formulated, and the dosage as well as the route of administration was altered.

The organomercurial diuretics were generally more effective than their nonmercurial predecessors, whereas the adverse reactions were infrequent, reversible, and rarely severe. When 900 patients were given a total of 5000 injections of Novasurol[R], 6% suffered from diarrhea, 4% from an irritation of the mouth, and a few vomitted or fainted (380). Other symptoms of organomercurial reactions were dizziness, fevers, chills, discolored skin, salivation, and an irritated colon. The medication also induced muscle weakness and spasms as well as shortness of breath or irregular respiration and such systemic reactions as declining blood pressure and heart palpitations (381). Multiple or premature heart contractions also caused minor heart injury in 10% of a group of patients checked with electrocardiographs. On rare occasions patients died after a fast, and irregular runaway heartbeat was triggered or a coma was induced by one or a series of injections, usually into veins. In some instances the patient died suddenly after the first injection (382).

Initially, the dosage prescribed depended on the desired diuretic response. A fluid loss of 3 to 5 lb was considered satisfactory, so less than that would mean a larger dose of the diuretic was prescribed for the next injection. Sometimes a total of 6 or 8 ml of a commercial preparation was injected in daily or twice daily doses. After several deaths 6 ml was set as the maximum nonlethal dose that an adult man could be expected to tolerate. Since children occasionally died after being injected with 1 or 2 ml, the dosage was gradually scaled down to the requirements of individual patients. As larger doses more frequently induced toxic reactions, the usual therapeutic dose came to be 40 to 80 mg of the mercurial in a 1- to 2-ml commercial preparation (383). A diuresis of from 6 to 8 liters of fluid could then be expected instead of the normal excretion of about 2 liters of fluid per day. To determine if the patient was unusually sensitive to the mercury compound, the initial test dose was often smaller, from 0.05 to 0.1 ml. An adverse response would signal the physician to extend the time between doses, decrease their size, change the route of administration, or substitute an alternative compound. As a last resort, this would be a less effective nonmercurial (384–385).

Since fatalities followed some injections into the veins, administration into the muscles or under the skin was preferred to slow the assimilation, and other routes were also tested. Rectal diuretics were quickly discarded because of unpredictable absorption rates and frequent irritation from mercurial suppositories. Tablets were most convenient, but they sometimes induced the old oral symptoms of mercury poisoning: a metallic taste, irritated gums, stomach pains, and an irritated colon. Only 5 to 10% was absorbed from the dose that was swallowed. Most of the oral diuretic was quickly excreted. However, less than 4 Neohydrin[R] or Cumertilin[R] tablets taken daily were the equivalent of approximately 40 to 48 mg of mercury and induced a diuresis equivalent to other preparations in doses of from 75 to 180 mg of mercury per day (383). The tablets were less often recommended for pregnant women after 1958 when one aborted her baby after swallowing 5 tablets at once (369).

Injections under the skin or in the muscles caused some gastrointestinal disorders, cramps, and nausea as well as local skin irritation, pain, and tender nodules, even after theophylline was added to mercurial solutions to diminish pain and irritation at the site of injection (381). A new diuretic, Thiomerin[R], was expected to cause fewer adverse reactions, because it contained only 39 mg Hg/ml and induced a diuresis equivalent to other such products. However, 42 out of 109 patients experienced temporary local pain or tenderness at the site of injection. This ranged from mild to intense and lasted from a minute to over an hour. Eventually, over 25% of the patients had troublesome local reactions to Thiomerin[R]. It also caused less heart but more kidney problems than some of its predecessors (385). A large and rapid diuresis could also induce leg cramps. While being treated for rheumatic heart disease, one 51-year-old man had cramps, vomiting, and extensive skin irritation. The platelets in his blood were also reduced over 50%, and his blood clot retraction time was prolonged. Because of negative reactions to Thiomerin[R], in the early 1950s Dicurin Procaine[R] was introduced in hopes that some individuals could better tolerate its procaine base. However, both induced similar diuretic effects and local skin irritation (386).

Because sensitivity to mercury sometimes increased over the course of several treatments, physicians also became more observant of warning signals. For instance, in 1931 a four-year-old child was given Salyrgan[R] intravenously in doses 1 week apart. An hour after the fourth injection he had a chill and temperature of 102°F. The next morning his skin had a blotchy rash that resembled measles, but his temperature had returned to normal. The rash was gone in 2 days, and a fifth injection was then administered. The child instantly coughed, cried out, and died (381). Other patients seemed to desensitize over time, so they might not be able to tolerate a mercurial on one occasion but later responded favorably to the same or a related compound.

Traditional symptoms of mercury poisoning were readily identified when the patient suffered loose teeth, skin rash, irritated veins and nerves or, rarely, a change in bone composition. Other reactions were unexpected (387). Fatalities that followed doses of Mercupurin[R] were initially attributed to "speed shock" from too rapid administration of the drug, but the blood clotted, whereas it normally did not in speed shock (388). To prevent this, the dosage or timing was changed.

A too rapid or profuse diuresis by itself could leave the patient prostrate with a weak pulse and lowered blood pressure. Another side effect, salt depletion, could cause weakness, nausea, drowsiness or restlessness, and mental confusion. A lack of salt would also decrease the effect of the subsequent doses of the diuretic. But the symptoms were not always clear-cut. A potassium deficiency could also cause muscular weakness and cardiac irregularities (383). To prevent the excessive loss of sodium, Mercuhydrin[R] was marketed in 1947.

Either mercury poisoning or salt depletion could cause emotional disorders that were particularly difficult to characterize among elderly heart patients who might change because of illness and aging as well as the medication. Since such subjective reactions are seldom directly traceable to the diuretics, these compounds may have caused more emotional and intellectual variations than were generally recognized. One 67-year-old man became very weak and then noisy when he suffered delusions of persecution after losing 30 lb on a regimen that included sedatives, dehydration through salt and fluid restriction, and two intramuscular injections of Mercupurin[R] in 1- and 2-ml doses. An old woman suffered more clearly defined personality changes each time she received a sequence of three injections of Mercupurin[R]. She became drowsy and uncooperative on three separate occasions and returned to normal afterwards (389).

As such side effects were documented by the mid1950s, many of the older organomercurial diuretics were replaced. Their successors corrected some adverse reactions while they generated others. Overall, the new mercurials caused the patients less discomfort and induced a diuresis that could be more carefully controlled. However, as older diuretics had been discontinued for patients with kidney damage after the site of diuresis was narrowed to that organ in 1928, some of the new ones also caused kidney damage (383, 390). This was difficult to detect if the kidneys were already defective. However, in 1958 five patients suffered kidney failure after extended treatment with mercurial diuretics, although their kidneys had been healthy initially. The three fatal cases were autopsied, and all had similar pathological changes that resembled acute mercuric chloride poisoning. Two victims had excessive mercury in the renal tubules. It was apparent that if albumin were not observed in the urine, kidney damage from mercurial diuretic therapy could be mistaken for a complication of the heart condition. The diuretics had to be administered to persons who could release both the excess fluid and the diuretic before the kidneys concentrated damaging mercury levels (390). Diuresis is now thought to occur when the walls of the renal tubules concentrate enough mercury to prevent the reabsorption of sodium ions and fluid (391). Then from 40 to 60% of the mercurial compounds are excreted within 2 days and nearly all within 3 to 8 days, although microscopic quantities may remain in the body, perhaps to concentrate in the brain.

Inorganic mercury and water-soluble organomercurial molecules can be distinguished with an isotope exchange technique. Thus it is possible to determine if the diuretic passes through the body intact or decomposes into inorganic mercury if it remains in the kidneys because of slow excretion or repeated dosage. But it is not known whether the organomercurial or inorganic mercury causes the diuresis. This information would help predict the rate of diuresis and the likelihood of kidney damage from the speed with which the compound decomposes. Studies on rats and mice indicate that the organomercurial diuretics decompose into inorganic mercury at varying rates in the same animals as well as in different species of animals. Damage to the kidneys of mice has been correlated with the extent of conversion from organomercurials into inorganic mercury (392). Moreover, experiments on tissues and urine specimens have sometimes made it possible to predict the potential diuretic capability of a mercurial compound before it was tested on human beings (390, 393–394).

Diuretics may add small quantities of mercury to the total body burden that concentrates in the brain, especially if human beings also methylate inorganic mercury biologically (98, 112,114). A positive spin-off application of organomercurial diuretics took advantage of their ability to cross the blood-brain barrier to improve techniques for spotting brain tumors. First, 1 ml of Mercuhydrin[R], a diuretic containing 39 mg of mercury, was injected to block further mercury uptake in the kidneys and reduce their likely exposure to radiation. Five hours later, 10 μg of Neohydrin[R] were administered as a radioactive mercury-203 compound. It concentrated in the brain tumor long enough to locate it with an X-ray before passing on through the body (395). This process gave a better picture of the tumor than the radioactive iodine-131 albumin that the mercury replaced. Nonetheless, because of their disadvantages, organomercurial diuretics have increasingly been replaced as satisfactory substitutes have become available.

ACRODYNIA: MERCURY AS A CHILDREN'S DISEASE

Children tend to be more susceptible to mercury than adults, and those who are hypersensitive may have adverse reactions to very small applications. However, mercury has been so common in nonprescription medications that the real cause of the allergic reactions was long unidentified and classed as a separate illness called acrodynia, also identified as Feer's disease or "pink disease" because it turned children's hands and feet a bluish pink. Adults were also long thought to be afflicted as the symptoms of several illnesses were lumped under one general designation, acrodynia. Physicians disagreed on which symptoms were essential to confirm the diagnosis as this catchall illness was variously considered an infectious (probably from a virus), postinfectious, allergic, endocrine, or toxic disorder. The cause was also theorized to be some kind of deficiency, neurosis, or a disease of the gastrointestinal tract (396).

In France in 1829, an epidemic of arsenic poisoning among adults was diagnosed as acrodynia (397–398). At other times the symptoms were considered to be not only pink hands and feet but scarlet cheeks and nose, peeling skin, loss of hair, teeth, and nails as well as salivation, excessive perspiration, a short-term rash, low blood pressure, itching, a burning sensation, and severe pain of the extremities. Other common symptoms were an increased pulse rate, extrasensitivity to light, and insomnia that often would accompany alternating moods of apathy and extreme irritability. Some of these symptoms are commonly associated with mercurialism but were not recognized in children.

Casual applications of calomel and other mercurial preparations stem from a long medical tradition often based more on historical precedent than scientific evaluation. The Chinese manufactured mercurous chloride at Hankow for centuries, and Indian chemists prepared it as early as the twelfth century. Like the Indians the Japanese also prepared calomel by mixing common salt, brick dust, alum, Indian aloe, and mercury and then brewing the mixture in a closed earthen pot for 3 days. After Maynerd, physician to Henry IV of France, introduced calomel in that country in the sixteenth century, Parisian physicians prepared a purer compound by a secret process that was finally published in 1608 (241).

Calomel was subsequently used to treat syphilis, as a laxative, and briefly as a diuretic before being discontinued in the nineteenth century because of negative side effects. In the United States mercurous chloride became almost a universal panacea. Medical textbooks of the last century list this all-purpose drug as a primary cure for such diverse illnesses as fever, diarrhea, heart disease, worms, rheumatism, and eye diseases (399). As everyone was "physicked," literally thousands of pounds of mercurous chloride passed through human beings, most of it without known damage whether or not it cured the particular ailment. The drug was so casually dispensed that it is not surprising that acrodynia was commonly diagnosed among adults before these medications were discontinued. But children were still treated with calomel and other mercurial compounds in ointments, teething powders, eyedrops, cold tablets, worm pills, soaps, dressings, washes, bath products, and disinfectants for cuts and bruises, as well as in diaper rinses. These nonprescription medications are still common enough so parents do not even think to mention them when a sick child is taken to the hospital, and physicians are not apt to ask about them either.

As some of the adult applications of mercury passed out of fashion, twentieth century pediatricians might have benefitted from the experience of some of their nineteenth century predecessors who were better acquainted with the horrors that could be wrought by mercurial preparations. They feared that mercurous chloride could convert into mercuric chloride (corrosive sublimate) if allowed to remain in the gastrointestinal tract for any length of time. Consequently, saline laxatives or castor oil were also prescribed to assure that the mercurous chloride was quickly eliminated. In 1865 Underwood observed that fatal gangrene developed in the mouths of several children after small amounts of mild mercurous chloride were administered (400). The compound was considered particularly dangerous for patients with obstructions of the bowels or intestines, irritated kidneys, advanced pulmonary tuberculosis, anemia, or alcoholism (401). The likelihood that calomel could convert into mercuric chloride when it came in contact with acid digestive juices was again noted in 1969. Then calomel cathartics were no longer permitted with an alkali iodide or an oxidizing or reducing agent, and a saline cathartic was recommended to follow the calomel within 5 to 8 hours. Before that, calomel had long been regarded as a relatively harmless cathartic, although both prescription and nonprescription mercurial drugs that were either consumed or absorbed had induced reactions that ranged from skin rash to gangrene of the mouth. Besides the bluish, swollen hands and feet, infants also developed digestive and vascomotor disturbances, multiple arthritis, insomnia, muscular weakness, and multiple abnormalities of the nervous system (402).

Sometimes customary doses of mercurial diuretics would cause children's mouths to become sore and their teeth to loosen. If the treatments were continued, the external ear could also be affected, and as other symptoms appeared, instead of identifying the diuretic as the cause, the ailment was diagnosed as the advanced stages of acrodynia. The central nervous system and kidneys could be damaged or mercurial gangrene could cause a devastating deterioration of the extremities that left the survivors with mutilated hands and feet and sometimes destroyed the lower half of the face.

Emotional reactions were not always identified in children either. Temper tantrums were attributed to normal childish behavior as those who suffered from mercurialism became irritable, hostile to their surroundings, and exceptionally prone to whine and complain. They would frequently appear depressed and might fly into a sudden rage. As the disease progressed, older children had insomnia, deliria, and hallucinations. They might either become so restless and excitable that they remained awake for days or weeks or they could be drowsy and apathetic all the time (396).

Despite the obvious symptoms, several factors kept mercurialism from being identified in children. The disease tended to be localized in some geographical areas and among certain families. In the southern part of the United States, for example, children are more often given worm pills than in other localities. And some families are more inclined to administer teething powders, skin ointments, cathartics, and other nonprescription medications. Since more than one family member would often contract the same symptoms, it appeared that the allergy might also tend to be a family characteristic passed on from parent to child. On the other hand, if a sensitivity response were built up over time, a child might suddenly become ill after tolerating compounds such as worm candies on previous occasions. The cause would not be suspected if he reacted negatively to what had been acceptable before. A delayed reaction could also be caused if the mercury were deposited in the body in an insoluble form and later dissolved to cause mercury intoxication. Then the time lapse between exposure and illness would often keep the two from being linked.

By the 1940s the symptoms of acrodynia were most often observed in children under 5 years old. Between 1923 and 1947 when deaths were tabulated in Great Britain, 63.5% of the victims were under 1 year old and mostly in the second 6 months of life. Of the remainder, 33.2% of the deaths occurred between the ages of one and two with about equal numbers of males and females. The deaths declined significantly during World War II and rose sharply in 1947. They occurred at the ages when children generally were teething and were less frequent when teething powders were not available. These teething compounds often contained calomel. When 20 children with acrodynia were checked in 1948, eighteen of them had high levels of mercury in their urine. Similarly, among 41 children mercury appeared in urine specimens from 38 or 92.7% (401). Among the control group 91.7% were free of mercury or showed negligible amounts in their urine (396).

Although these tests indicated that acrodynia was really mercury intoxication, attempts failed to correlate mercury levels in urine with the severity of the disease. It is now known that the mercury is more apt to be excreted in the feces. However, only urine tests were conducted in 1948, and they made documentation difficult. One child had little mercury in the first urine sample but continued to excrete it for 9 months. Others only infrequently excreted mercury. Before he died, one child excreted very little mercury, but an autopsy showed that large amounts were stored in his kidneys.

Calomel teething powders continued to be dispensed after they were linked with mercurialism in 1948, and acrodynia still was occasionally improperly diagnosed. A 19-month-old child was treated for albumin in the urine and excess fluid for a month before mercury poisoning was diagnosed. Before correct treatment could be instituted, his condition deteriorated until he died. When the cause of illness was improperly diagnosed the first time, a hypersensitive child could sometimes suffer repeated disasters. One 5-month-old boy survived pink disease only to die of kidney damage when he was again given a teething powder a year later (403).

When acrodynia was finally linked with nonprescription medications administered to children, the cause of illness was more frequently identified by carefully quizzing the parents about their use of home remedies. One mother inadvertently volunteered that she had been in the habit of giving her child a worm candy about once a week over the preceding year. Another case was solved 3 years after a 14-month-old girl had originally been admitted to the hospital when the mother recalled that the child had been given "cold tablets" that contained mercury (396). Mercurialism was similarly induced in a 3-year-old girl when her mother gave her some gray tablets to cure "liver trouble." Since these 0.5-g tablets each contained 23 mg of mercury, a total dose of approximately 15 g was administered over 5 months (402). After a correct diagnosis of mercury poisoning, treatment with dimercaprol usually helped expedite excretion if the kidneys or central nervous system had not been damaged (404). However, unusual hypersensitivity continues to produce illnesses from other mercury applications. For instance, a 1-year-old boy died after exposure to diapers that had been rinsed in a mercuric chloride solution. These extreme reactions are most unusual, and many of the compounds are being replaced by nonmercurials. However, since these nonprescription medications continue to be readily available, parents should be better informed about their potential hazards and not dispense them casually (396).

16

CONTINUING APPLICATIONS

SKIN PREPARATIONS AND ANTISEPTICS

Although mercury is no longer used to treat syphilis and the quantity in diuretics decreased from 7600 lb in 1967 to 5700 lb by 1973 and continues to decline, in that year the pharmaceutical industry still consumed 46,057 lb of mercury (405). Both prescription and nonprescription ointments and antiseptics with high percentages of mercury continue to be widely applied to the human body as well as other creatures with some therapeutic benefits and occasional negative side effects. In addition, although most mercurials were banned from cosmetics and phenylmercuric acetate is no longer used on toothbrushes, this mercury compound is still added to face creams to retard the growth of bacteria. And in tattoos the red dye is made from cinnabar. Therefore, small quantities of the element are absorbed from these sources and may remain as part of the total body burden with the other input from air, water, and food.

Ointments

Where mercury is concerned, the line between health and beauty aids can be drawn on intent in many cases and is very thin in all cases. Until 1973 a primary distinction was that the ingredients were listed on a jar of cream used to cure skin rashes but were not on a jar of face cream. Presently, nonprescription mercury ointments may contain 30% mercury whereas prescription ointments may be 70% elemental mercury embedded in an oil base. Since Aristotle originally recommended that the element be diluted with saliva before being applied to the skin, elemental mercury and its compounds have been incorporated in innumerable red, white, blue, and yellow ointments used to treat many skin and eye irritations.

Red mercuric oxide ointment was a longtime favorite to treat eye irritations before and during the first part of the nineteenth century, although the sharp cornered crystals would sometimes damage the surface of the eye. Therefore, a smoother yellow mercuric oxide compound was formulated in 1856. Ten years later, a British doctor named Pagenstrecher treated a few specific eye diseases with individually formulated prescriptions for which he added mercuric chloride to a solution of potash and then mixed the precipitated oxide into a white ointment base (406). Before the end of the nineteenth century, however, a standardized yellow mercuric oxide ointment in a petrolatum base was being freely dispensed for home use as well as in medical schools and clinics to cure such diverse eye ailments as inflamed outer and inner eyelids, corneal ulcers, stys, and inflamed eye lashes. Not only could this relatively insoluble compound irritate and scar the cornea, but it also might prevent the eye from healing and thereby prolong recovery. Thus medical opinion was divided on the value of this compound before 1933 when a slit lamp revealed granules of the substance in a patient's pupil. Although petrolatum alone is as effective to soften the crust that forms on the eyelid, the ointment continued to be prescribed for nodules on the inner eyelid and cornea as well as for irritated mucous membranes. As physicians debated the value of yellow oxide ointment, druggists continued to dispense it, and over 5500 lb of mercury still went into its manufacture in 1948.

Applied over several years, the 1% yellow ointment could discolor the eyelids or cause changes in the eye itself. Compounds with a higher percentage of mercury accelerated these changes. In 1946 after a woman applied a 6.6% ammoniated mercury ointment to her eyelids, they were filled with bluish granules that formed a dark pigment around the eye (407). The ointment's prolonged use caused such gradual discoloring that the individual was not aware of it until a physician pointed it out during treatment for another medical condition (408). Another eye treatment, ammoniated mercury opthalmic ointment, usually contained only 3%

ammoniated mercury and generated less severe side effects than a mercuric cyanide compound applied to relieve irritated eyes. It sometimes caused heart flutter, excessive perspiration, albuminuria, swelling and redness of the hands, insomnia, and emotional disorders (396).

Similarly, a series of mercury ointments used to treat other skin conditions have also generated negative symptoms. A 1% yellow mercuric oxide has been popular to treat itching, scaling, and pubic lice. A 5% white ammoniated mercury ointment has been applied for pinworms, pubic lice, and scabies caused by mites that burrow beneath the skin. Now pubic lice are more often treated with nonmercurials, but the 5% ammoniated mercury ointments continue to be applied to treat such varied skin conditions as impetigo, psoriasis, eczema, and anal itching (409–410).

Mercury ointments that react chemically with the proteins and enzymes in the skin tissues may irritate and cause a skin rash that usually clears up when the medication is discontinued. The many mercury ointments sold without a prescription and applied without a doctor's supervision have generally caused only mild, infrequent skin reactions. But on a mercury-sensitive individual, the small rashes can become weeping sores. In 1883 a 55-year-old woman sought medical attention because the chronic eczema on her feet induced an annoying itch, and her skin had become rough and discolored over many months of treatment. When an ointment containing mercury and a compound of tincture of lavender and vaseline was prescribed to cure the rash, it turned into running sores that spread to her eyebrows and cheeks (411).

Children's diaper rash and other skin irritations may also be intensified by mercury ointments. When a 20-month-old boy had a single blemish treated with ammoniated mercury ointment, a rash spread over his entire body. It was tentatively diagnosed as German measles until his skin peeled and his swollen and distended eyelids left only slits to see through. Recovery was rapid after the ointment was discontinued and olive oil applied (412). Because such severe reactions have been infrequent, nonprescription mercury ointments have continued to be marketed, and parents sometimes have applied them freely and injudiciously. In 1958 when a baby boy was being treated for a threadworm infection, his jealous 3-year-old sister insisted that she be given the same medication. The mother obliged, and both children were admitted to the hospital suffering from mercurialism at the same time (413).

The familiar blue ointment containing 25% elemental mercury in petrolatum has been applied to every imaginable skin condition on man and beast, and some animals, notably horses, have shown an unusual sensitivity to mercurials. A 7-year-old Percheron died of uncontrolled bleeding attributed to mercury in the bloodstream after approximately 25 g of red mercuric iodide ointment were applied to cure a lame fetlock joint. After a dog was treated for a cyst on its leg, it licked off the mercury ointment and developed a sore mouth and peeling feet (414). And an upper New York state veterinarian humorously described the misuse of a traditional remedy to rid cattle of lice that old-timers passed along from one generation to another. Originally, the remedy called for boring a small hole in the stanchion and filling it with blue ointment. Then a wooden plug was inserted. Over the years, the blue ointment came to be placed directly on the stanchion and animal's hide. This had caused a skin irritation on a 2-month-old calf that was intensified when the calf licked and rubbed itself. A young bull in the next pen also licked the calf and developed an irritation on its legs. The veterinarian remarked that he saw similar cases every few years, and the cows generally developed mild skin lesions but "no fatalities to either cows or lice" (415).

Cosmetics: Creams and Tattoos

In 1968 skin preparations consumed 15,200 lb of mercury. Of this, 3800 lb were for therapeutic applications such as skin and eye ointments. This use was expected to decline since drug manufacturers were under pressure to replace mercury with more effective and less harmful substitutes. However, cosmetic manufacturers were not required to list their ingredients or even to pretest their products or report negative reactions. They were expected to add 50% more mercury to their products by 1970 and to increase the total to 21,000 lb in 1973 (416). However, in that year a new law was finally passed requiring that the ingredients in cosmetics be listed on the containers as they were with medications. Until then, trade secrets were often protected at the expense of the consumers. In addition, all mercurials were banned from cosmetics except phenylmercuric acetate, which prevents the growth of bacteria in face creams. Manufacturers received this concession because they contended that no effective substitute was available.

Unfortunately, this exception means that a long-documented and highly visible negative reaction to mercurials has been allowed to continue. The victims of this skin discoloration have most often been women who applied face creams for a number of years. Such incidents have occasionally been described in the medical literature since 1920. A school teacher had students and co-workers imply that she did not wash her face properly, because her skin had turned a bluish gray that looked dirty. Under magnification her skin appeared to be covered with minute black dots caused by 15-year's application of a well-known brand of face cream that contained mercury and bismuth (417). Another woman developed such conventional symptoms of mercury poisoning as sensitive gums and loose teeth as well as darkened skin around her eyes and on her throat after repeated applications of a compound called Gourand's Oriental Cream that turned out to be a suspension of calomel in water. Although these women would not permit a sample of skin to be taken, in 1925 another patient permitted a biopsy that showed mercury extending through several layers of skin in a true pigmentation that could not be removed with skin lighteners. Although this condition was not uncommon, it was sometimes misdiagnosed and thought to be the result of fatigue, cardiac problems, hepatitis, or a chronic disease.

The condition was reversible in some instances. After a 42-year-old cosmetician's skin had been discolored by a massage cream called "Derma-tone," much of the shaded appearance disappeared after the tiny globules of elemental mercury were treated with a tincture of iodine to form a soluble mercuric iodide that dissolved in alcohol (418–419). Although all these incidents occurred before 1930, the problem continued to recur as the products went unregulated and the clients confirmed both the short- and long-term effects by tests on their own bodies. Thus in 1965 a 79-year-old woman displayed facial discoloration from applying a 4% ammoniated mercury night cream for more than 50 years (420).

Although mercurials in face creams have darkened the skin, in Nairobi, Kenya, skin lighteners containing mercury were more apt to cause urinary problems among women. Such creams had been marketed in Kenya for 15 years before 1972 when four young nurses contracted the nephrotic syndrome after applying compounds that contained from 5 to 10% amino-mercuric chloride (421). The condition cleared up when they stopped using these products.

Whereas skin creams are apt to cause afflictions among women, the tattoo is more often associated with the youthful exuberance of male sailors on shore leave. Tattoo artists use red dye made from cinnabar and needles that engrave dots of color under the skin to weave hearts and flowers with girls' names, tributes to mother, or references to one's national origin. These embellishments usually cause little consternation beyond a wife's annoyance over the name of a girlfriend or a nude form that might shock the congregation if a sleeve were rolled too high on a warm day. But in unsanitary tattoo parlors, a syphilitic tattoo artist can infect his clients with an unsterilized needle or an already infected person could localize and accent the secondary eruption of this disease where the skin was irritated by a tattoo. However, this hazard may be lessened somewhat if the cinnabar acts as an antiseptic when the needles are dipped into the red dye (422). Moreover, in some instances syphilis appears to have been locally inhibited in the red part of the tattoo (423).

On the other hand, mercury-sensitive individuals may suffer a reaction to the cinnabar when the tattoo is implanted initially or after additional exposures to mercury even after several months or years. One tattoo remained dormant for 9 years until a minor cut in the middle of the design precipitated a skin rash that spread over the man's entire body. It did not clear up until the tattoo was completely removed and replaced by a skin graft (424). Sometimes applying another mercury compound such as a skin ointment can cause a localized flare-up in the tattoo (425). One individual could induce a rash on his arm simply by rubbing a red rose tattoo. It disappeared when the rose was left alone.

Even if implanting the tattoo caused no initial skin irritation, subsequent mercury exposures sometimes built to a sensitivity reaction. One man was tattooed as a youth and treated for syphilis with mercury ointment at age 43 without ill effects. But at 60 when he was treated for hemorrhoids, the red part of his tattoo began to swell, itch, and ooze. The rash disappeared when the medication was stopped, but a patch test with ammoniated mercury ointment confirmed that he had become allergic to the element (425). Other individuals have experienced allergic reactions when dentists performed spot tattoos with cinnabar to locate and mark the hinge axis to study the patient's bite (426–427). This sensitivity response could be compounded if the teeth were filled with mercury amalgams.

Antiseptics

Despite the superiority of other antiseptics, nonprescription mercurials like Mercurochrome[R] and Merthiolate[R] have gained such wide popular acceptance that they continue to be sold over the counter to disinfect the skin and mucous membranes as well as to prevent infections in small injuries such as cuts and bruises. Nonetheless, the effect of mercurial antiseptics on microorganisms is still not completely understood. One widely held theory is that both inorganic and organic mercury compounds contain ions that inhibit the sulfhydryl-containing enzymes of the microbes. In this case the enzymes' activity is restored when the heavy metal is removed or highly diluted. If the mercury is absorbed, it might deactivate the necessary sulfhydryl enzymes in the body as well as in the bacteria on the skin (409).

Mercuric salts became popular antiseptics after Robert Koch (1843–1910) conducted experiments with anthrax spores in 1881. Even greatly diluted mercuric chloride solutions seemed to be powerful germicides that killed the most resistant microorganisms. Later, mercuric chloride was shown to only restrain microorganisms from multiplying, and this effect is seriously diminished in the presence of organic matter. Moreover, the compound coagulates protein and irritates the skin if it is applied over long periods of time. Injections under the skin can also cause toxic reactions (428). Nonetheless, mercurial antiseptics continued in wide use, probably because of the influence of early researchers like Koch and also for lack of an adequate substitute.

On a cut, Mercurochrome[R] can cause the bacteria to become dormant, but the germs are only suppressed, not destroyed. Not only is the effectiveness of the compound reduced by organic residues on the skin such as dirt and acid body fluids, but the germs may continue to grow and reproduce when the bacteriostat is washed off (428). Mercurochrome[R] is usually marketed as a 2% solution containing alcohol, acetone, and water or in bulk as a powder (389). It may contain up to 25% mercury, and children are more susceptible than adults to apparently harmless therapeutic doses. After a 3-month-old baby girl had sores in her mouth swabbed with a 1000 ppb solution of Merthiolate[R] from 4 to 6 times a day for 5 weeks, her condition gradually deteriorated until she was left moderately retarded after mercury poisoning was finally diagnosed (429).

Nor did the organomercurial compounds live up to expectations, although they were widely promoted as reliable antiseptics for cold sterilization of surgical and dental instruments after World War I. Through 1937 they were recommended in medical journals, textbooks, and other standard references. The Council of Pharmacy and Chemistry of the American Medical Association also accepted the claim that one organomercurial compound was a potent germicide against spores and spore-bearing bacteria.

In 1937 when all the dental schools and many practicing dentists in the United States were surveyed, 71% of the respondents advocated chemical agents to sterilize instruments such as cutting blades, scalpels, nerve broaches, tissue shears, and instruments made of rubber or other substances that would be damaged by boiling. Sixty-three percent favored a mercurial and listed such products as: Metaphen^R, Mercresin^R, Mercarbolide^R, and Merthiolate^R. About 50% of the dentists and dental schools favored a 400-ppm solution of Metaphen^R (428).

Despite their popularity, the effectiveness of these organomercurials had only been tested on a few organisms and no anaerobic spores. When extensive tests were conducted on the antiseptics and disinfectants listed in the 1937 edition of *New and Nonofficial Remedies*, none of the mercurials could be depended on to sterilize instruments within 24 hours, even in the absence of deterrents such as organic matter. When the spores of *Clostridium welchii*, *Clostridium tetani*, and *Clostridium sporogenes* were tested in mercurial solutions containing from 170 to 20,000 ppm in a pH ranging from 3.5 to 10.5, the commercial preparations with the highest and lowest acidity were most active, but none of them killed the organisms. Even under optimal conditions, the best the mercurials could do was to prevent the spores from causing infection in some cases (428). Moreover, among 52 sutures suspended in a solution of 1000-ppm potassium mercuric iodide for more than ten years, 12 still contained viable spores, seven of which were anaerobic bacteria. Similarly, among 120 knives used in surgical operations, 12% were still contaminated with spores of anaerobic bacteria after cold sterilization.

By 1940 evidence showed that the feeble mercurial bacteriostats were not effective to sterilize surgical and dental instruments. However, a 1000-ppm solution of mercuric chloride, the oldest mercury antiseptic, has continued to be used to sterilize surgical instruments that are harmed by boiling (429). And 500-ppm solutions sometimes have also been employed as a surgeons' handwash, although the compound is harsh on the skin and corrodes metal instruments. Until recently, a phenylmercuric compound was also included in an antibacterial fabric softener to control *Staphylococcus* in hospital laundries. It also could cause a skin rash on hypersensitive individuals, but this and similar compounds were banned in 1970 primarily to prevent their release into the waterways (430).

DENTISTRY

Dentists continue to favor mercury amalgam fillings, because they remain a soft plastic mass while being shaped in the tooth cavity and then harden into strong, abrasion-resistant, relatively permanent replacements for decaying teeth. Usually the alloy consists of 26% tin and 74% silver with enough mercury to make it pliable while being shaped. The typical proportions by weight are 5 parts alloy to 8 of mercury. Many alloys contain tin, copper, and sometimes zinc (431). From an average of about 2000 flasks per year in 1960, in 1972 dental preparations required 2983 flasks of mercury or 5.6% of the total consumed in the United States. This declined to 2679 flasks in 1973, and the projected decline in population growth makes it likely that this mercury application will also expand more slowly in the future (270, 293).

The mercury alloy generally replaced gold which was easier to work with but more expensive than other metals. A number of other alloys had been tested previously. One called "fusable metal" consisted of bismuth, lead, tin, and mercury. Its melting point was just under the 212°F at which water boils. The melted mixture was poured into the patient's tooth cavity through a small funnel, presumably while his head was hanging down if an upper tooth were being filled. This was replaced by a combination of metals that melted at 140°F. Small solid pieces could be placed in the tooth and touched with hot instruments to melt and fill the tooth. Burns to mouth and face then were hazards of dentistry (432).

In 1826 M. Taveau, a Parisian dentist, used powder filed from silver coins and mixed with mercury to form a soft putty amalgam that he called silver paste. When this compound was placed in the tooth cavity, it hardened as the silver powder and mercury condensed. The inexpensive compound did not require heating, but the filling sometimes expanded and split the crown of the tooth or swelled until the bite was irregular and induced pain. However, the state of the art soon advanced to better alloys and improved procedures. The technique of filling teeth with mercury amalgams was reportedly brought to the United States from France by two brothers named Crawcour in the 1830s, and it touched off what has since been called "the Great Amalgam War" between dentists who thought the new amalgams were superior, at least for their less affluent patients, and those who still preferred gold.

Leonard J. Goldwater, a noted mercury scholar, contended that many dentists were professionally antagonistic toward mercury amalgams because they had invested heavily in the gold trade. For whatever reason, they vigorously attempted to discredit mercury amalgams by contending that the poisonous element could endanger the patients' health (236). When the American Society of Dental Surgeons was founded in 1840, inserting amalgams was declared a malpractice. The members signed a pledge that amalgams were unsuitable to fill teeth or fangs. Those who refused either resigned or were drummed out of the association. Despite this organized resistance, by the end of the nineteenth century the proportions had been standardized for a durable silver amalgam that is pliable initially because of mercury. Most teeth are still filled with this substance.

During their brief exposure, even the most hypersensitive patients are not likely to suffer an adverse reaction, because the amalgam fillings generally harden quickly (433–434). However, dentists and their assistants absorb mercury through their hands and inhale more from vapors emitted as fresh amalgams are inserted or the old ones are drilled out. The daily exposure intensifies if sanitary precautions are lax, and waste mercury accumulates where it has been dropped or spilled on the floor.

Dentists and especially their assistants often have not really been aware of the potential harm from mercury "with the resulting almost total absence of even the most elementary precautionary measures" (435). Since the small droplets of mercury are not readily visible and the vapors are colorless and odorless, the hazard is not apparent. However, in the early 1970s the problem was again stressed in Canadian schools where dentists and their assistants were trained (274). More safety precautions were recommended after the levels of mercury were checked in dental personnel and their offices and laboratories.

Some vapors are continually exhausted from new applications and previous deposits of mercury in the rooms. In a number of dental laboratories, waiting rooms, and offices in the Canadian province of Saskatchewan, the mercury background levels were lowest early in the morning and highest when dental work was in progress. Whether the rooms were relatively new or over 20 years old, the time factor made little difference. Floor coverings were more important. When the rooms were carpeted and automatic mixers were operated, the mercury vapor levels were frequently excessive. The Canadian 0.05 mg Hg/m³ threshold limit value (TLV) was exceeded in 10 out of 27 carpeted dental laboratories with flooring an average of 3 years old. The level was below the TLV in a tiled dental laboratory, whereas it was elevated in the dentist's carpeted office, apparently because mercury had been tracked from one room to the other and then vaporized out of the carpet (274). Similarly, mercury levels may be elevated in the living quarters when dental offices are in private residences.

While dental work is in progress, the vapor levels often elevate briefly but dramatically. When old fillings were removed, 0.4 mg Hg/m³ were measured for two minutes, and 0.18 mg Hg/m³ were recorded in another instance (435–436). In 1968 when the American TLV was 0.1 mg Hg/m³, 50 offices ranged from 10% of that limit when the office opened in the morning to 400% while the dentist milled out a silver alloy filling and carved a replacement (437).

Mercury is deposited in the carpeting and cracks in the floor as the vapor condenses and also when the element is spattered or carelessly discarded. Hand mulling has long been discouraged for sanitation as well as health as this 1935 address to an association of dentists indicates.

I find a definite lymphocytosis in the great majority of dentists that I have examined. These cases clear up only on elimination of all sources, I mean not only to stop mixing mercury in the palm of the hands and then wipe away with your fingers the excess amount which overflows the cavity, not only to scrub and wash from your hands the minutest traces of mercury before eating, but particularly to stop throwing mercury on the floor, for I believe that the prolonged inhalation of mercurilized dust is a far more potent factor (437).

However, hand mulling continues. Thus mercury is absorbed through the dentist's skin, and the amalgam is less apt to adhere well in the patient's tooth cavity because of the accumulated dirt, oil, and perspiration on the dentist's hands (438).

In this process the correct portions of mercury, tin, and silver are rolled together to amalgamate just before the tooth is filled. Then the pliable mixture has to be shaped quickly before it hardens. Although excess mercury prolongs the hardening time, it is often carelessly measured and then squeezed out and discarded during amalgamation, often on the floor or in an open container. To prevent exposure to vapors during mulling, some technicians squeeze the amalgam in a cloth instead of in their hands or wear rubber gloves and cover their faces with a surgical mask to filter out particles (439).

Equipment is also available to dispense and combine the elements mechanically, although these "jiggle bugs" may also spatter mercury around. However, none is lost from prepackaged capsules with the correct portions of the metals in separate compartments ready to be combined and mulled when pressure is applied. Then the amalgam does not have to be touched until the tooth is filled. Although the packages are sanitary and easy to store and use, they are also more expensive and so not as common, nor are the new alloys with an inert inner core that does not react with mercury. These compounds must be prepared in a ball mill.

Amalgam powder is also expelled when new fillings are ground and polished or the old fillings are removed. Ultrasonic equipment can produce a cloud of minute mercury droplets and alloy particles (440–441). During high-speed drilling this can be minimized and the heat reduced by operating a water coolant. Otherwise, unless the flooring is free of cracks, the mercury droplets accumulate and are difficult to remove. Disposing of waste mercury in a container covered with a layer of bacteriostatic solution prevents vaporization as does air evacuation equipment to vent the mercury vapors outdoors. Air conditioners can also break up the droplets and recirculate the vapors where they can be inhaled.

When adequate safety precautions were disregarded, 90% of the dental personnel had from 10 to 155 ppb of mercury in their urine, whereas 80% of a control group had none (437). More mercury was also recorded in the hair, toenails, and fingernails of 20 Canadian female dental assistants than in other citizens. The assistants' fingernails averaged 68.8 ppm, whereas their hair ranged from 12.0 to 45.0 ppm. More mercury kept concentrating in their hair while the control group maintained a relatively stable range from 0.2 to 4.0 ppm in 1962 and 0.2 to 6.0 ppm in 1969–1970 (442–443). Their toenail samples had a mean value of 5.1 ppm, whereas the surgical assistants averaged 9.3 ppm.

Generally, dental personnel did not concentrate enough mercury to cause acute symptoms of poisoning such as hand tremors that would force a dentist out of practice. However, the chronic exposure could be responsible for subclinical reactions of a more subtle nature such as depression, headaches, fatigue, or excitability (436). Although they are difficult to pinpoint, mercurialism was confirmed when a dentist was referred to an eye specialist because he complained of headaches and fatigue.

As the lens of his eyes was clear, his vision had remained normal, although a 3 to 3.5 mm metallic disc had formed from minute deposits of mercury in his pupils. The small grayish-blue circles of the disc had several faint rays extending out at the edges like a "sunflower" cataract. This type of mercurialentis had previously been observed among industrial workers exposed to mercury or in people who treated their eyes or eyelids with mercury compounds. Spectrographic analyses of the dentist's urine, feces, and blood also revealed 33 ppm mercury and smaller quantities of other elements normally contained in dental amalgams (444–445).

Dental patients were more apt to experience adverse reactions to the earlier copper-mercury amalgam fillings, as Dr. Alfred Stock observed in 1926. These amalgams were more soluble and released more mercury than modern alloys (446). As billions of fillings were inserted in teeth, the patients usually had slightly elevated urinary mercury levels only for from 1 to 4 days (439). Thirty-seven dental patients concentrated a maximum of 9 ppb mercury in urine 4 days after they had very large fillings implanted. When 114 people were checked before and after having teeth filled, six had slightly over 0.5 ppm mercury in the urine to begin with and slightly more the day after the tooth was filled. Hypersensitive individuals who experienced allergic reactions to mercury in amalgam fillings usually already had an elevated body burden. Then the small addition caused a skin rash and sore mouth. However, occasional sensitivity reactions are more intense (447).

A 32-year-old woman had averaged two or three fillings on each regular visit to the dentist since she was 20 years old, and she had 12 silver amalgam fillings implanted without complication in 1957. But when she was pregnant in 1960 and had two silver amalgam fillings inserted under local anesthesia, she developed a skin rash on her chin and neck. A few days later when more amalgam fillings were inserted, the skin rash increased and spread to her armpits and the insides of her arms. More dental work was then postponed for 11 months. However, a few hours after two more silver amalgam fillings were inserted, a more acute skin rash returned with itching, blisters, and oozing. When it healed, patch and scratch tests proved the patient was allergic to both freshly prepared and

2-week-old silver amalgam. Her sensitivity intensified with each exposure, and the reactions were generally more pronounced on the side of her body where the fillings were inserted or tests had been performed (442).

Such rare sensitivity responses to amalgam fillings may be more readily isolated than problems with teeth caused by mercury poisoning in other circumstances. One employee of a thermometer factory had his painful gums and loose teeth diagnosed as pyorrhea. After all his teeth had been pulled, the man developed other symptoms of mercurialism such as a skin rash and tremors (298). Such incorrect diagnoses and unnecessary dental work are more likely than mercury poisoning from the dental amalgams or exposure during insertion.

IV

CONCLUSION

17

PREVENTION OR CURE: HUMAN POISONING

Killing oneself or others, either accidentally or intentionally, has always been a challenge that reflected popular means available at the time. Today, barbituates, motor vehicles, and guns have replaced the mercury compounds, especially mercuric chloride (corrosive sublimate), that previously were conveniently available in powders and tablets to be readily converted into death-dealing overdoses. Observations of accidental poisoning and suicide attempts established the symptoms of mercury poisoning and afforded the opportunity to test various remedies, mostly useless, before BAL was discovered in 1940.

White precipitate (ammoniated mercury) was a popular ingredient in skin ointments dispensed by London pharmacies at least by 1746. If the ointments were consumed, the injury was thought to be caused by the ½ to 3% mercuric chloride in the compound. Even in 1850 when white precipitate contained up to 70% mercury, chemists and toxicologists did not consider the elemental form harmful. However, poisonings by ammoniated mercury or corrosive sublimate were relatively common in the nineteenth century. Children were more severely afflicted than adults who more often lived to describe their mistake. However, there were many exceptions. Because of the low literacy rate, people who could not read the label only belatedly realized that they had the wrong medication, even when it was clearly designated as poisonous. Incidents like the following were not uncommon.

The patient was a strong young woman, deaf and dumb, and unable to read. Her mother, a very stupid old woman, had asked at the druggists', according to her own account, for a penny seidlitz powder (a cathartic preparation), and was asked by the assistant who served her whether she wanted it *white* or *red*. The daughter swallowed the whole penny-worth (about 2½ to 3 grams); and it was only then, soon after she began to suffer from vomiting and severe pain in the stomach, that some of the neighbors discovered that the packet was labelled "white precipitate—poison" (448).

As people accidentally swallowed white precipitate instead of milk of sulfur, that is, precipitated sulfur, medical practitioners were gradually able to determine the dose response in human beings (449). When a 48-year-old woman was given white precipitate instead of a preparation ordered for her neuralgic pains, the precise quantity and time of administration could be charted accurately. After she consumed 20 grains (1.3 g) of the powder, she exhibited conventional symptoms of mercury poisoning and recovered in 6 weeks (450).

Murder attempts generally did not elicit precise dose response information, and the evidence often could not be legally validated, because the patients excreted it before they died of exhaustion from the sickness. However, they also gave physicians the opportunity to dispense such up-to-the-minute medications as frequent doses of egg along with five drops of opium extract every 5 minutes as well as ten drops of witch hazel extract every quarter of an hour. This treatment was not sufficient to save a 52-year-old baker who had consumed approximately 40 grains of white precipitate (451).

After a New York City newspaper described what was believed to be a fast and painless death in 1911, the blue, coffin-shaped 0.5-g mercuric chloride tablets enjoyed a burst of popularity among the suicide prone. The tablets were widely marketed as household antiseptics, to treat infections in family members, to rid the house of bed vermin, and to prevent conception. However, the newspaper account had grossly underestimated the pain (225).

After the pills were swallowed, the corrosive mercuric chloride burned the mouth and gastrointestinal tract so severely that death or shock could be anticipated within 48 hours. The initial pain usually propelled the individual to the hospital where the only treatment available was to swallow great quantities of liquid, often milk and egg whites, to prevent corrosion of the gastrointestinal tract. If the patient survived this first phase, according to a prevalent medical slogan, he "virtually lives or dies by his kidneys." Within 4 days the kidneys often failed and caused death from acute uremia. If the victim survived this, he could contract gangrene of the colon within 12 days as the mercury collected to be excreted in the feces. Nothing could be done to relieve the kidney malfunction, and removing the colon to prevent gangrene was considered too drastic a remedy (226). Therefore, the usual prescription was to drink liquids and await the outcome. One such poisoning occurred at the hospital when a surgical patient was accidentally injected with 1.5 grains (about 98 mg) of mercuric chloride salt during a postoperative saline infusion. Since 0.5 grain was usually fatal, medical practitioners observed this illness from the initial vomiting through kidney obstruction to colonitis and death (225).

A more exotic array of treatments was offered to a 27-year-old World War I veteran who attempted to kill himself with tablets that he thought were silver nitrate. Instead, they were the familiar mercuric chloride tablets. By 1925 they were used only to poison rats, bedbugs, and other vermin as well as to treat flypaper. The veteran died 2 weeks after the onset of acute mercury poisoning despite a number of ineffectual treatments that included catherization, enemas of soapsuds, morphine with mustard and warm water, and strychnine. The patient refused to take olive oil, but he later accepted bismuth and paregoric as well as strychnine sulfate. After his throat became irritated, probably from the medication, an alkaline antiseptic gargle was also prescribed. This patient's experience led the attending physicians to again recommend that acute mercury poisoning be treated by having the patient consume egg whites and highly diluted milk simultaneously. Then they recommended an emetic to remove the albuminous mercury compound from the stomach (226).

After decades of ineffectual remedies ranging from brandy and beef tea to egg whites and opium, an organic chelating agent was finally developed in the early 1940s that could at least expedite the excretion of inorganic mercury compounds before they could do as much harm. BAL was originally formulated in 1940 with research undertaken to combat arsenic poisoning. Since Lewisite [dichloro (2-chlorovinyl) arsine], arsenical mustard gas, had inflicted great injury among troops engaged in trench warfare during World War I, antidotes were quickly sought when World War II began in 1939. Lewisite and other arsenicals blistered the skin because the arsenoxides blocked the sulfhydryl components of the tissue protein enzymes so the carbohydrates could not oxidize. Sodium arsenite and Lewisite acted selectively to raise the blood's pyruvate level, whereas fat-soluble chloroarsines penetrated the keratin layer of the skin to destroy the cells of the epidermis, dermis, and capillaries.

Since antiseptics that normally killed bacteria could be neutralized by adding sulfhydryl compounds, it seemed likely that other living cells might also be protected by whatever saved the bacteria. When cysteine, thioglycollic acid, or glutathione was added, bacteria survived such anti-spirochetal compounds as arsenoxide, arsphenamine, neoarsphenamine, mercury, and bismuth (426). Whereas traces of mercuric chloride generally retarded the growth of *Escherichia coli*, these bacteria grew normally if a little glutathione and more thiolacetic acid were added (427).

Since the skin became irritated in contact with thioarsenites because the arsenic combined with the tissue thiols, a compound was sought to prevent this by forming a more stable single-dithiol ring. Simple molecular 1:2 and 1:3 dithiols such as toluene dithiol and ethane dithiol were not effective, but dimercaprol (2,3-dimercapto-l-propanol) in a penetrating oil could protect the skin almost completely against the 50% inhibition of enzymatic activity caused by Lewisite. Moreover, the toxic action could be reversed after exposure. A rabbit's skin could be saved 20 minutes after it was contaminated, and human volunteers with skin irritation and swelling did not blister when treated an hour after exposure to Lewisite and phenyldichloroarsine. Whether injected or applied to the skin, the new compound halted the toxic action of Lewisite on the pyruvate oxidase system 2 hours after rats and guinea pigs were contaminated and already showing signs of illness.

For security reasons during the war, this potent antidote for gas warfare was code named OX 217, but the Americans called it British anti-Lewisite (BAL) in recognition of the British scientists who discovered 2,3-dimercapto-l-propanol (452). Although not reported in scientific journals until the war ended, by 1943 physicians on both sides of the Atlantic Ocean were using BAL to treat victims of arsenic poisoning in factories or those who suffered complications from arsenic therapy. At the Edgewood Arsenal workers who were accidentally exposed to poison gases like adamsite or diphenylamine chloroarsine were sent to Johns Hopkins Hospital for BAL ointments or injections to counteract the arsenical dermatitis (453). When treatments for syphilis caused dermatitis and encephalitis, BAL could force more rapid excretion of the arsenic. The skin rash usually cleared up within 2 weeks, depending on the individual's sensitivity and level of exposure (454).

Then BAL also was tested and proved to be more effective than rongalite (sodium formaldehyde sulfoxylate) as an antidote for mercuric chloride poisoning. Rongalite could reduce the mortality from mercuric chloride poisoning in cats if treatment were not delayed more than 15 minutes (445). The need for haste made this an unlikely treatment for people, although a 5% solution of rongalite could reduce the mercuric chloride to metallic mercury which was not as readily absorbed and so was excreted from the human body (456). In tests on pigeon brain brei, six molecular proportions of BAL completely protected the enzyme system from 50 ppm mercuric chloride, a concentration that normally would inhibit 50% of the unprotected enzyme. Even half as much BAL afforded a significant degree of protection. When rats were injected with mercuric chloride and then BAL, acute mercury poisoning was prevented. The antidote could also be combined in injections and oral doses. In one study 39 out of 40 untreated rats and 9 treated with rongalite died, whereas all 12 rats treated with BAL survived (457).

As BAL was more effective than rongalite, monothiols were an improvement over dithiols. Rabbits injected with varying doses of mercuric chloride were saved from systemic mercury poisoning if BAL thiosorbitol or BAL glucoside were administered quickly. BAL glucoside prevented kidney damage, whereas BAL alone did not, but the treatment had to be fast. The rabbits suffered kidney damage in 30 minutes and died in 60. After a delay of up to 5 hours, however, dogs could be treated with doses of BAL that could be tolerated by human beings, and BAL glucoside was even more effective. The mercury bound to BAL glucoside in a less toxic combination that dissociated to a greater extent than mercury bound to BAL.

Although BAL glucoside was more effective, it was not available in a sufficiently pure form to warrant clinical trial on human beings, so BAL was used instead (458). Tests were conducted at Johns Hopkins Hospital which received many patients who had attempted suicide with mercuric chloride as well as the victims of accidental exposure to poison gases at the Edgewood Arsenal. Among 42 people who had swallowed from 0.5 to 20 g of mercuric chloride, many were desperately ill when they entered the hospital, but only two died after BAL treatments. Previously, at this hospital 27 out of 86 poisoning victims died after swallowing 1.0 g or more of mercuric chloride even though treatment was started within 4 hours of ingestion. Prompt treatment with BAL saved persons who had consumed from 3 to 20 g of the poison (456, 459). During and immediately after World War II, only adults were treated with BAL. Later, children suffering from acrodynia also received the compound and improved substantially faster than untreated children (460).

Although BAL hastened the excretion of mercury, it also induced some negative side effects. A few patients complained that their tongue, lips, or extremities tingled temporarily after a single injection. The symptoms usually appeared within a few minutes and climaxed in less than half an hour. Like mercury intoxication, BAL could cause nausea and vomiting, headaches, tremors, and shakiness (453). BAL caused severe pain of the gums, lips, and lower extremities when two college students were treated for exposure to vapors emitted while heating hearing aid batteries to retrieve mercuric oxide. The type of mercury compound and the patient's general condition influenced the reaction to BAL. Many complications could be prevented by injecting 50 mg of ephedrine sulfate half an hour before treatment (461). Usually BAL did the most good when administered within 4 hours after exposure, and it seemed to release more mercury from alkaline than acid urine. If renal damage had already occurred when the BAL was administered, death from uremia was less likely if hemodialysis were also given (462–463).

Like BAL, *D*-penicillamine also expedited the release of mercury from the body. When this compound was first tested on rats, they were completely protected after being injected with 3 mg of mercuric chloride per kg of body weight at the static dose level (464). After the success of these animal experiments prompted tests on human beings, *n*-acetyl-*d, l*-penicillamine benefitted both a gilder who was finally striken with mercurialism after years of chronic exposure without ill effects and a 5-year-old boy who was exposed to mercury vapors in a freshly painted bedroom and perhaps by eating some of the paint (219, 423). However, this compound could also cause adverse reactions among those who were allergic to penicillin, and none of the treatments were as effective against chronic mercurialism as they were for acute exposure.

Quick treatment with BAL can often prevent damage in acute poisoning where the kidney is the critical organ, but in chronic mercurialism the brain accumulates methylmercury and only very slowly eliminates it. When the metal has penetrated the cellular membrane, BAL does not bind with it as easily and in some cases seems to accelerate its passage into the brain. As the cells and central nervous system are destroyed, this affliction has no cure (465–466). At the same time, prevention is becoming steadily more difficult. Although people can consume less fish with high mercury levels if they have a choice of foods, some of the element is still concentrated from food and from the atmosphere. To halt this process will require substantial changes in the way waste discharges are handled. Although the methods are costly, the element can be reclaimed from industrial proc-esses, but no satisfactory methods are yet available to remove the deposits of mercury that are already lodged in the waterways.

18

MERCURY REMOVAL
FROM THE WATERWAYS

Since many waterways had accumulated substantial quantities of waste mercury in the water column and sediments by the time elevated levels were discovered in fish, research was initiated to find methods of retrieving the element as well as preventing further degradation of the waterways. The 1972 Water Quality Amendments banned the discharge of more waste into the waterways after 1985, so the first priority was to find methods of purifying industrial effluents, ideally by operating a closed recycling system where the same water could be reused instead of being released into the waterways. However, as a temporary measure to decrease mercury discharges in 1970, settling ponds were quickly built or the waste was pumped into handy wells. But the retaining ponds overflowed, and wastes that percolated through wells might contaminate the underground aquifers from which drinking water is drawn. Thus experiments were conducted to devise permanent methods of removal.

Although it is chemically and sometimes biologically feasible to remove mercury from factory drainage or settling ponds, the tons that already contaminate the natural waterways present removal problems for which no satisfactory solution has been devised. The cheapest and most popular tactic is to hope the natural currents will eventually cover the mercury-laden sediments and flush the mercury bound to suspended particles into the oceans. Estimates of how long this would take range from 10 to 1000 years. In the meantime less than 0.1% of the total mercury in the aquatic ecosystem is generally converted into methylmercury each year (105). The conversion rate depends on the quantity available as well as other variables in individual waterways, so it is difficult to predict how much mercury the aquatic organisms will concentrate.

If the mercury in the sediments is covered up, it may be prevented from methylating. To do this, natural sedimentation is the cheapest and consequently the preferred method. The process has been charted in Pickwick Lake since 1970. By 1972 from 5 to 7 years was the time projected to accumulate a layer of 1.5 to 2 in. of sediment at the current rate (467). A faster method would be to put a layer of sand or gravel over the sediments. At least 2 in. were estimated to be necessary to cover the mercury adequately and prevent penetration by small aquatic creatures. In 1972 fifty acres could be covered for an estimated $18,500, although costs would vary depending on inflation, how much contaminated sediment had to be covered, how far the sand had to be hauled, and how deep the water was. Another factor was whether the covering would be permanent or if water currents would be apt to disrupt it (468). Another negative side effect could be the loss of macrofauna if the limnic ecosystem were altered (29,103).

Instead of sand and gravel, inert clays and freshly ground silicate could also bind with the mercury to provide a stable covering. Ground silicate binds with metallic and inorganic ionic mercury more readily than with phenylmercury. Sludge from ore concentration plants also contains heavy metals that form strong bonds with mercuric cations and monomethylmercury cations (48, 91, 105). In a lake treated with this sludge a preliminary survey yielded fish with less than 0.01 ppm of mercury, whereas those from untreated control lakes measured 0.9 to 1.2 ppm. Iron turnings also bind with mercury in solution or in sediment (469). Therefore, metal such as crushed automobile bodies could also cover and bind with the mercury. However, the layer of automobiles would have to be covered with sand to prevent seepage. In laboratory experiments a layer of 2 to 3 cm of fluorspar tailings also reduced the formation of methylmercury by about 80%. However, natural turbulence or macroinvertebrates could disturb this covering and release the mercury into the waterway (113).

In experimental tanks several types of polymer film overlays were also tested such as high-density polyethylene, low-density polyethylene, polyvinyl chloride, alcohol soluble nylon, and poly(ethylene-vinyl acetate).

Preformed nylon 6 (polycaprolactam) was the most effective barrier against both organic and inorganic mercury. Copolymer hot melts also could control mercuric chloride, but none was as impermeable to the organic mercury compounds as polycaprolactam (470).

The preformed plastic films can be rolled out and pinned on the bottom of the watercourses, but installation has the major disadvantage that a large bulk of materials would have to be handled under water. Hot-melt processes could also be adapted to individual waterways. The disadvantage here is that the dispensing equipment would be expensive to develop, construct, and maintain. Coagulable polymers also are very costly. However, a uniform film of about 0.012 in. has been formed as alcohol was replaced by water when the nylon polymer was extruded into the watercourse. The polymer films adhered to the bottom reasonably well in tests, but in actual waterways covered with debris, they might not attach to the uneven surface. Moreover, the expensive film could be torn or displaced by currents. All the films require some form of anchoring and holes that permit gases to escape from the sediments. Besides being expensive and temporary, the films would also destroy the benthic ecosystem (313).

Since the buried mercury could be uncovered despite layers of sand, metal, or plastic, other alternatives are to bind it chemically, remove it, or dike it off. Chemicals in leaching fluids as well as on mesh, plastics, and brass have been tested to bind with mercury permanently or temporarily for removal from the waterways. The long-chain alkyl thiols, inorganic sulfide, or natural proteins that can remove mercury from waste effluent before the factory discharge enters the waterway also have some application for waste reclamation, although the problems vary in the natural environment. For one thing, methods that work in laboratory experiments or where a limited quantity of waste has to be treated may be prohibitively expensive even to apply to high concentrations of sediments close to factory outfalls.

Thiols, some of the most effective mercury-binding agents, are oily liquids that may impart an objectionable taste or odor to fish as they prevent the uptake of mercury. Since they float on water, the thiols would also have to be sunk to the sediments, perhaps by absorbing them on hydrophobic oil-sinking agents. Although the thiols are costly, only a small quantity would be required. An estimated 200 ppm of thiol in sediments would be sufficient to bind 100 ppm of mercury effectively. In 1972 mercaptans ranged from a high of 80 to a low of 30¢ per lb or from $100 to $244 per acre to treat the sediments with *n*-dodecyl mercaptan (471).

A number of natural substances have also been tested to bind with the mercury, but they raise new waste disposal and mercury reclamation problems. Peat, sawdust, and peanut hulls have been proposed as binding agents as well as hair and feathers with sulfides added. At 4¢ per lb, feathers appeared to be cheap, but they had to be condensed; thus approximately one ton of feathers is needed to complex a pound of mercury. Therefore, a layer of condensed feathers 4.5 in. thick would cost at least $2880, and if they were not removed, the feathers would have to be covered with approximately 825 tons of sand to maintain a ½-in. thickness. Moreover, the feathers would degrade and release the mercury again (471–472).

Fibers offer more potential as mercury scavengers. Nylon, cotton, and wool have been tested. Depending on the circumstances, nylon removed only 20 to 50% of the mercury, whereas wool would reclaim from 90 to 95% of the inorganic or organic mercury within 24 hours. Although felt is particularly absorbent, a layer 0.13 to 0.5 in. thick on an acre of sediment could cost from $1500 to $12,000 depending on the thickness of the layer. Fifteen thousand pounds of wool would remove an estimated 4500 lb of mercury. Cotton fibers that were chemically modified with nitrogen compounds could also extract mercury from an aqueous solution. In tests on cotton impregnated with cross-linked polyethylenimine (CPEI), 0-(2-diethylamine)ethyl (DEAE), and 0-(2-aminoethyl) (AE), the fibers removed up to 3000 ppb of mercury from the solution. After the mercury was extracted with sulfur-coated cotton net, it could be reclaimed from anaerobic sediments. Chopped cotton could also remove mercury and then be regenerated to use again (473).

Some gels also hold mercury long enough to retrieve it. A 25- to 50-ppm polyvinyl alcohol gel containing sulfur or phenyl thiourea can remove inorganic mercury or methylmercuric chloride. Elemental mercury can be removed by elemental sulfur or thiourea on a paraffin base (474). After the mercury adheres to the mesh and binds with the sulfur in the gel, it can be removed from the water to be disposed of or reclaimed. All these processes are expensive, and some remove only part of the mercury. They can also cause additional waste removal problems. Therefore, dredging would seem to be the most obvious way to remove mercury-laden sediments, but it has also turned out to be one of the most controversial methods.

Harbors and channels in the Great Lakes were dredged for over 100 years to keep them open for navigation by lake and ocean vessels. At least until the mid1960s the Corps of Engineers and the Public Health Service had not regarded dredging as a cause of pollution in the Great Lakes (475). When methylation raised a new specter in 1970, dredging operations were hastily halted. Until then, a procession of hopper dredges, barges, and scows dredged from 80,000 to 190,000 cubic feet of sediment out of the shifting bed of the St. Clair River, the cutoff, and Lake St. Clair to be dumped in Lake Erie each year. Throughout the Great Lakes about 10.8 million cubic yards of sediment were regularly dredged to keep 115 harbors open, and the residues were deposited in about 100 disposal areas of the lakes. They were designated to avoid interfering with water intakes or beaches. Each year, the rivers then carried more solids into the navigation channels and gradually filled them up again.

On the route from the St. Lawrence Seaway to Chicago or Duluth, ocean-going vessels passed through the St. Clair River system from Lake Erie to Lake Huron and churned up sediments with their propellors. By 1970 the St. Clair River had shoaled badly from its dredged level of 28 feet below the low-water datum. Then the impact of continued dredging had to be studied again from an economic standpoint. Environmentalists had pressured the Corps of Engineers and the Department of the Interior to examine the ecological effects of dredging and disposal in the 1960s. In 1965 the Public Health Service had reported that municipalities and industries caused most of the pollution in waterways by discharging inadequately treated sewage and industrial wastes (475).

Dredging was assumed to be harmless because the natural currents would carry the solids into the lakes anyway, and the man-made procedures only expedited the process. In the late 1960s the Corps of Engineers and the Department of the Interior conducted a 2-year investigation that cost $8 million and ascertained that dredging brought an estimated 17% of the suspended solids and 8% of the total solids into Lake Erie. A smaller percentage of the solids would be derived from dredging lakes where this operation was not as prevalent. The proportion of organic pollutants in the dredging spoils depended on the condition of the harbor. Normally, organic pollutants were estimated at less than 10% of the total, but at Cleveland they were about half (475).

Even if it does not stir up mercury to be methylated, dredging out the sediments is not possible where the sides of the waterway rise precipitously or the water is too deep. Even where maintaining navigation seems essential, as in the St. Clair channel and harbors on Lake Erie, new EPA standards also affect dredging alternatives. Sediments with more than 1 ppm of mercury are classified as so contaminated that they should be discharged only in the ocean at depths greater than 100 fathoms or on land disposal sites (471). Thus contaminated sediments in the St. Clair waterway have to be diked off or disposed of on land. But the Corps of Engineers met with strong opposition from environmentalists and the Walpole Indians when they proposed to dike off a disposal area on Dickinson Island. The Indians feared that both their lands and fishing water would then be contaminated.

When dredging is feasible, new techniques have to be devised that stir up as few sediments as possible and prevent the mercury from being released back into the waterway. Barriers placed around mechanical dredges can limit the dispersal of mercury into the water column. But water drawn out with suction dredges is apt to retain or drain back up to 50% of the mercury. In some instances only 2% of the mercury remained in the sediments (475). Instead of pulling it out of the waterway, one possible alternative is a sinkable barge with watertight cofferdams to pump the water out of the impoundment temporarily while the mercury in the sediments is dissolved with a liquid hypochloride leach that could then be removed. A 1% hypochloride solution with a pH between 6.0 and 6.5 removed the mercury in laboratory experiments but not in field tests in San Francisco Bay where the element was firmly bound to sediments with a high organic content (471). Another possibility is to treat drainage water and dredging spoils with lime, aluminum, or iron to bind the mercury and keep it from draining back into the waterways (476).

Aside from dredging and open-lake dumping, dikes may be the least costly method of removal, and they can offer some advantages. In the early 1970s open-lake dumping cost in the range of 78¢ per cubic yard, whereas dikes cost from 3 to 9 times more, depending on land use and other factors. The least expensive alternative treatment costs $5.11 per yard. At Astabula, Ohio, open-lake disposal cost 35¢, whereas other treatment cost a minimum of $9.08 per yard, a 26-fold increase. The Secretary of the Army estimated that 41 harbors with polluted sediments could be diked off at an initial cost of $112 million over 2 or 3 years with an annual $5 million in operating costs (475). Besides their expense, landfill operations could be unpopular if they were not sealed over to prevent the release of noxious gases. Landfills also raise the possibility that drinking water supplies will be contaminated if the pollutants percolate into the underground aquifers. However, an advantage in some cases is that the reclaimed land could be converted into valuable waterfront parks. Cleveland officials have also proposed that dredging spoils be used to build an offshore island where an international airport could be located. All these plans are expensive, some are highly impractical, and without continued public pressure many people question if the benefits of reclamation are worth the cost.

19

MERCURY AND THE LAW

EARLY REGULATIONS AND INITIAL LAWSUITS

Although mercury contamination was a new public issue in the spring of 1970, water pollution and navigation rights were very old hat. In fact, the present American Constitution grew out of a convention called initially to discuss navigation rights on the Chesapeake Bay. And in 1886 Congress directed the U.S. Army Corps of Engineers to prohibit the dumping of refuse into New York Harbor (477). This statute originally was intended to safeguard navigation. However, British law set a precedent for water pollution control in 1892 when a Scottish distillery sued a mine for polluting a river from which the distillery drew its water. The judge ruled that the property owner on the bank of the river or stream "is entitled to the water of his stream in its natural flow without sensible diminution or increase and without sensible alteration in its character" (478).

The American Congress subsequently passed the Rivers and Harbors Act, also identified as the Refuse Act of 1899 (479). Navigable interstate waterways were to be protected from solid refuse contamination that would impede or obstruct navigation. Water and other liquid refuse that flowed from streets and sewers were specifically exempted. Timothy Atkeson, General Council for the Council on Environmental Quality, described the 1899 Refuse Act as "that sparkling innovation in anti-pollution legislation of the McKinley administration" (480), because 70 years later it proved

233

to be the most readily enforceable anti-pollution law on the books because of two simply worded, no-nonsense provisions. Section 13 forbade dumping wastes into navigable waterways without a permit from the Corps of Engineers, and Section 16 proclaimed that violations were misdemeanors punishable by fines of no less than $500 or more than $2500 per day or by imprisonment for not less than 30 days or more than 1 year. But the law had not been enforced. By 1970 the Corps of Engineers had only issued 415 permits, most of them after 1965.

Federal laws passed after 1889 more often authorized the treatment of wastewater than preventing further contamination. In 1912 the Public Health Service was authorized to investigate the health effects of pollutants in navigable lakes and streams, but officials were generally not given authority to institute corrective measures (481). However, after the Oil Pollution Act of 1924 prohibited dumping oil into navigable coastal waters, the Secretary of War was charged with enforcement (482). In 1948 the Federal Water Protection and Control Act gave the states the primary rights and responsibilities for the control of water pollution. Funds were allocated to states, municipalities, and interstate agencies for research and waste treatment facilities. A federal court suit could be instituted to obtain pollution abatement in interstate waters only after the states consented, the discharger was notified twice, and a public hearing was held (483). The procedures were slow and cumbersome under this temporary measure initially passed for 5 years and then extended for 3 more until the first permanent water pollution control act was passed in 1956. Although state programs then were still encouraged with federal grants, the new three-step enforcement procedure did not require state consent for implementation. Instead, interstate water pollution could be halted after a federal-state conference, a public hearing, and a federal court suit (484).

After the Fish and Wildlife Coordination Act became law in 1958, the Corps of Engineers was required to consider water quality in terms of the environmental impact on fish and wildlife when licensing refuse discharges (485). Although the corps requested that dischargers' permit applications list the possible environmental impact of waste effluent, this mattered little, since most companies did not apply for permits. Moreover, mercury was not among the toxic substances listed on the few permits that were issued. After 1960 when Justice William O. Douglas ruled in the Republic Steel Case that the 1899 Refuse Act could be applied to prevent environmental contaminants as well as navigation hazards, the Corps of Engineers steadfastly remained the overseer of navigation hazards but not water quality. Nonetheless, lower courts subsequently supported Douglas' interpretation in other lawsuits.

The 1961 Federal Water Quality Control Act extended and strengthened federal authority over intrastate and coastal navigable waters, and the federal water pollution control program was transferred to the Department of Health, Education and Welfare (486). However, more central administration was still needed to coordinate state and federal water quality efforts. Therefore, the Water Quality Act of 1965 placed all federal water pollution control programs under the newly created Federal Water Pollution Control Administration. This act also provided for the establishment, revision, and enforcement of water quality standards for navigable interstate waters (487). The states were given a timetable in which to create their own water quality standards and submit the plan for federal approval. More funds were also allocated for research and to construct treatment facilities for wastewater and land disposal sites to dispose of solid waste (488). The states were given more incentives to adopt water quality standards under the 1966 Clean Water Restoration Act. Among these were grants for projects to improve methods of waste treatment and water purification (489). The Federal Water Quality Control Administration (FWQA) was also transferred again, this time to the Department of the Interior. In 1970 it came under the EPA at the height of the mercury contamination controversy when old enforcement codes were finally being tested.

Secretary of the Interior Walter Hickel invoked the cumbersome water quality enforcement procedures against the Aniline Film Corporation of New Jersey in July 1970. The company quickly agreed to limit its waste discharges during the 6-month warning period, and this case was not pursued. However, it was apparent that a full cycle of hearings, lawsuits, and appeals would make enforcement a very slow procedure (490).

Hickel had acted when public concern about water quality intensified after the discovery of elevated mercury levels in fish in American waterways. When Senator Philip Hart originally began an investigation into mercury contamination in the Senate Subcommittee on Energy, Natural Resources and the Environment on May 8, 1970, the first hearings were held in Mt. Clemens, Michigan, because mercury pollution was thought to be a local problem in the St. Clair waterway. Representatives from the

Dow and Wyandotte Chemical Companies testified that although their firms had released mercury into the St. Clair and Detroit Rivers for many years, they had no idea that this could be harmful. State and federal health authorities seconded their claim and pleaded inadequate budget and manpower to maintain extensive monitoring programs.

When the Hart Subcommittee reconvened in Washington at the end of July, the U.S. Geological Survey and the FWQA had checked other waterways, and the FWQA reported mercury contamination in 20 states. Two of them, Alabama and Michigan, had banned fishing in contaminated waterways, and citizens were pressuring their congressmen for action. On July 9, 1970 Representative Rogers of Florida called on the Department of Health, Education and Welfare to halt the dumping of mercury into waterways immediately. He also wanted the Department of the Interior to review the 0.5-ppm action level for mercury in fish (491). Secretary of the Interior Hickel sent telegrams on July 14, 1970, pledging federal support to 17 states where mercury contamination had been confirmed. Hickel also promised that the administration was developing "hard evidence" and would seek court action in any confirmed case of mercury pollution "if corrective measures are not taken swiftly on local levels" (492). Nonetheless, Governor Albert Brewer of Alabama accused Hickel of "nothing but buck-passing," because the federal government promised only to make a study if the states did not. And the federal government did not appear too aggressive on other fronts either.

At the Lands and Resources Division of the Justice Department, Assistant Attorney General Shiro Kashiwa said it would not be in the genuine interest of the government to prosecute polluters who spent significant amounts of money trying to clean up (493). And the day after Representative Rogers called for action from the Department of the Interior, U.S. Attorneys around the country were issued guidelines for litigation under the Refuse Act that made prosecution highly unlikely.

The policy of the Department of Justice with respect to the enforcement of the Refuse Act for purposes other than the protection of the navigable capacity of our national waters, is not to attempt to use it as a pollution abatement statute in competition with the Federal Pollution Control Act or with the State Pollution abatement procedures, but rather to use it to punish the occasional or recalcitrant polluter, or to abate continuing sources of pollution, which for some reason or other have not been subjected to a proceeding conducted by the Federal Water Quality Administration or by a State, or where in the opinion of the Federal Water Quality Administration the polluter has failed to comply with the obligations under such a procedure . . . (494).

Representative Henry Reuss (D-Wisconsin) subsequently charged that the Department of Justice enforced the law against the occasional polluter but not against big corporations that continuously violate the pollution laws. He contended that this ragged enforcement "breeds contempt and disrespect for the law" (494). Assistant Attorney General Kashiwa, on the other hand, told the Conservation Foundation that the Water Quality Improvement Act, Section 10(c) was intended to prevent the federal government from licensing pollution declared unlawful by the states (495). In other words, the states were protected from a more liberal federal policy.

Kashiwa justified the new guidelines on the grounds that they advised the local U.S. Attorneys of what action could be taken under the Refuse Act, what federal agencies could assist them in securing proof of the allegations of discharges in violation of the Refuse Act, and to acquaint them with the fact that the department believed that actions not merely for fines and imprisonment but also for injunctive relief could, in appropriate cases, be brought under the Refuse Act (494). Despite the potential for criminal prosecution with fines and imprisonment, the Justice Department emphasized injunctive relief which would only require that the discharges be halted. The revised guidelines still required U.S. Attorneys to contact Washington before filing civil or criminal complaints or returning indictments in Refuse Act cases. In the House Subcommittee on Conservation and Natural Resources of the Committee on Government Operations, Reuss questioned why the U.S. Attorney should have to contact Washington before taking action against a polluter unless it was to give federal officials an opportunity to prevent it on political or other grounds. On August 14, 1970 Reuss told the House of Representatives that the Department of Justice totally abdicated its statutory duty to enforce the 1899 act vigorously, and he denounced Attorney General John Mitchell as a "scofflaw where water pollution is concerned" (494).

The Bass Anglers Society got the lawsuits moving. From its headquarters at Montgomery, Alabama, the organization's president, Ray Scott, first publicized John Mitchell's orders to the U.S. Attorneys not to enforce the 1899 Refuse Act and then instituted the largest environmental pollution suit in history. On July 22, 1970, in three U.S. District Courts in Alabama, class action suits were filed against 193 Alabama businesses and industries as well as 21 municipalities. All were charged with polluting streams (496). The Anglers also sued the Secretary of the Army and the Alabama Water Improvement Commission and called on the U.S. Army Corps of Engineers to set standards for refuse dumping (497). The Reuss Subcommittee was also pressuring the corps to enforce the permit program under the 1899 Refuse Act. And New York Representative Richard L. Ottinger, who had successfully sued the Penn Central Railroad for polluting the Hudson River, also sued the Justice Department to force action (496).

Two days after the Bass Anglers filed suit, Secretary Hickel gave the Justice Department the names of 13 companies illegally releasing mercury into the waterways and requested that they be prosecuted. The names of the companies were not revealed initially, and a spokesman at the Justice Department would only confirm that "we did receive some referrals from Interior" (498). The list had been narrowed to eight companies with 10 plants by the time the Justice Department began court proceedings on July 24, 1970. The eight were Georgia-Pacific at Bellingham, Washington; Olin Mathieson Chemical Company at Niagara Falls, New York, and Augusta, Georgia; Oxford Paper Company at Rumford, Maine; Weyerhauser at Long View, Washington; Diamond Shamrock at Delaware City, Delaware, and Muscle Shoals, Alabama; Allied Chemical Company at Solvay, New York; International Mining and Chemical Company's Chlor-Alkali Division at Orrington, Maine; and Penwalt Chemical Company at Calvert City, Kentucky. Hickel said these suits were just the beginning and others would follow, but they were both the beginning and the end of the lawsuits filed against mercury polluters at that time (499).

In the first place, civil instead of criminal suits were instituted to obtain abatement by injunction, a modest goal compared with suits by the Province of Ontario and the State of Ohio against the Dow Chemical Company of Sarnia to force the company to clean up the waterways and compensate commercial fishermen and tourist camp operators for their losses. These suits were still pending 7 years later, and none were instituted in American courts to force the companies to clean up the previous damage. In fact, even the abatement goal was quickly compromised. Whereas Secretary

Hickel had declared that no discharges at all would be permitted, Carl L. Klein, Assistant Secretary of the Interior for Water Quality and Research, told the Hart Subcommittee that it would take 6 months to reduce the mercury discharges as low as possible. As Klein emphasized that abatement was the immediate goal in 1970 and previous losses would be cleaned up later, he also reiterated that the 1899 Refuse Act was primarily a navigable waters act and that the primary authority to control pollution still belonged to the states (500).

Like Hickel, Klein refused to name the total list of 51 companies that were discharging mercury into the waterways illegally. Secrecy was generally the order of the day. Although Ralph Nader charged that two memos had been sent to personnel in the Department of the Interior warning them not to release any information, Klein denied the charge even as he refused to give the facts requested by the Hart Subcommittee. In contrast, Nader told the Hart Subcommittee that mercury pollution was a national disaster for which habitually polluting industrialists should be removed from their jobs and jailed, whereas some polluting firms should be forced into environmental bankruptcy (501).

Secretary Hickel finally released the names of all 51 polluters on September 16, 1970 as he announced that they had reduced their mercury discharges by 86%, from 287 to 40 lb of mercury per day (222, 502). Hickel praised the companies for their cooperation "when the safety of our nation is at stake" and reported that an additional 79 companies also used mercury but discharged no detectable quantities into the waterways (503–508). Nonetheless, Hickel was increasingly criticized for unnecessarily and unfairly whipping a few companies in public without adequate warning (505). He drew the most fire for acting prematurely in prosecuting the Georgia-Pacific Company at Bellingham, Washington. This company acknowledged dumping 41.5 lb of mercury daily on July 22, 1970, when the lawsuits were instituted. But the company soon was able to reduce this to 10.5 lb and then to 8 oz per day as they introduced a total recycling system that eventually eliminated even trace discharges to Puget Sound (506–507). Although he refused to apologize for a possible mistake in acting hastily against Georgia-Pacific when it had abatement facilities under development, Hickel sent a letter to company officials praising their new abatement program. Hickel subsequently resigned his post, and the FWQA was transferred to the EPA where many of the same officials continued essentially the same policies. For instance, when 18 more mercury polluters were discovered in a survey of 884 companies, the EPA also refused to release the names. John Quarles defended the secrecy policy to the Reuss Subcommittee on the grounds that the agency only announced categories of problems and the filing of abatement suits, but adding more polluters to the list did neither (508).

Nor was Quarles able to explain why only 13 mercury polluters had initially been singled out and eight prosecuted. He indicated that this action was hastily taken, because the mercury crisis had been portrayed as a national emergency. Nonetheless, the eight widely distributed chloralkali plants were not all major polluters. Quarles did say the states were given the first chance to prosecute, and some companies were not indicted if they acted quickly to clean up the pollution on their own. For instance, the Dow Chemical Company's facility at Plaquemine, Louisiana, had reduced its mercury losses from a range of 40 to 50 lb a day to 3.2 lb by May 18, 1970 (509). However, Quarles insisted that the EPA would use the courts to enforce pollution abatement if necessary, because "it does dramatize the problem and stimulate the public attention which frequently will break the log jam and achieve results" (508). By July 1971 EPA had referred 37 cases to the Justice Department for prosecution in criminal and civil suits and had provided advice and technical assistance in 175 instances of Refuse Act violations. When court cases resulted in judgments, the fines generally ranged from $500 to $1000, but the mercury polluters did not even get this slap on the hand.

After Hickel resigned as Secretary of the Interior, John Mitchell announced that the Justice Department had successfully resolved the mercury cases. Commerce Secretary Maurice Stans released a National Industrial Pollution Report stating that mercury discharges had almost totally ended within a few months, and the high concentrations from past discharges could be expected to disperse gradually to larger areas (510). So much for cleaning up the waterways! And the Justice Department refused to tell the Reuss Subcommittee how much mercury the polluters were still discharging, how it was being measured, or what agreements were being made with the companies to reduce their losses. All the subcommittee could learn about the 10 continuing suits against the eight firms was that the Justice Department had a long-standing policy not to disseminate information on pending litigation. Moreover, Justice Department officials contended that disclosure would violate Canon 20 of the American Bar Association Canons of Professional Ethics. Reuss retorted that companies could avoid disclosure by letting the Justice Department sue and then enter into a stipulation to furnish information if its discharges were kept confidential.

The day after William O. Douglas delivered the Supreme Court's concurring opinion in the Pentagon Papers Case that "secrecy in government is fundamentally anti-democratic, perpetuating bureaucratic errors," Reuss reconvened the subcommittee and called on John Quarles of EPA and Assistant Attorney General Kashiwa to provide the long-sought information on the 18 additional mercury polluting industries and what was happening in the civil and criminal cases filed under the Refuse Act (508). Despite the Supreme Court ruling, Kashiwa still refused to release information on the mercury suits without the consent of the eight companies (114). However, only one, Olin Mathieson, subsequently refused to release it (102, 511).

It turned out that the Justice Department had negotiated agreements that permitted the ten companies to continue releasing half a pound of mercury per day or 10 lb in a 21-day period, all at once or in increments (508). Most of the lawsuits were continued before a date for a hearing was set, but one was scheduled for the Oxford Paper Division of the Ethyl Corporation before company officials closed the plant (512). They announced that production could not be continued economically without mercury discharges. From the 26.7 lb of mercury previously discharged daily, the accumulation remains in the sediments of the Androscoggin River and Casco Bay (508). The other companies continued production, and John Quarles subsequently announced that they reduced their mercury discharges a total of 98.4% or from 139 to 2 lb per day (493). All the initial 51 industrial discharges were supposed to have reduced their mercury losses from 86 to 97% by April 1, 1971 (102, 493, 511, 513).

Despite the excellent reports of pollution abatement, some uncertainty remains about the accuracy of monitoring procedures and even about the particulars of the pollution abatement agreements with the companies. Abatement was achieved largely through verbal agreements, although Quarles indicated that the final form would eventually be "reduced to writing" and would vary with the individual companies.

I expect that there would be some forms of internal memorandums, but either you have a signed stipulation between the company and the government or you don't. And I think I have already indicated that in many of these cases—if not most of them, I believe that we do not have signed stipulations between the company and the government, but the company knows and we know what has been agreed upon (494).

On being informed that a Swedish chloralkali plant lost no mercury to the environment, Quarles indicated that feasibility was relative, and a no-discharge standard might be possible in some cases but not others.

Despite the federal hoopla over successes in mercury pollution abatement, some state officials contended that the companies could diminish their mercury losses further if adequate pressure were applied. R. S. Howard, Executive Secretary of the Georgia Water Quality Control Board, contended that the Olin Mathieson Corporation had losses of only 1 oz a day when the federal government authorized an 8-oz release. The furious Howard charged the Justice Department with "pulling the rug out from under me" (514). Moreover, while the mercury cases were being negotiated, the whole program of pollution abatement was being revised.

POLLUTION PERMITS

Prior to 1972

After Reuss' Subcommittee rediscovered the Refuse Act of 1899, they called on the U.S. Army Corps of Engineers to explain its permit program of the past 70 years. Of the 415 permits that had been issued, 266 were still in effect. None had ever been granted in 22 states including the highly industrialized Connecticut, Maryland, Michigan, Ohio, Rhode Island, and Virginia (493). Except in New Jersey, California, and Louisiana, less than 25 extant corps permits had been issued since January 1965. In a letter to Lieutenant General F. J. Clarke, Reuss called on the corps to enforce the law.

This meager number of existing Corps permits issued for the discharge of industrial waste is disgraceful, when one contemplates the numerous industries in each State that undoubtedly discharge pollutants into our waterways. The time has long passed for these industries to stop flouting the 1899 law and to either comply with it and the regulations issued thereunder or to cease discharging their wastes into our waterways (515).

Then the corps revised its regulations and requested state certification before permits would be authorized (516). After June 17, 1970 the states had to provide "reasonable assurance" that the applicants would not violate current water quality standards with their future waste discharges. If the states refused to act, they waived their right to participate. Either way, to be effective the Corps of Engineers had to enforce the 1899 act. Four days after Walter Hickel sent his list of 10 mercury polluters to the Justice Department, Reuss encouraged Clarke to publicize the 1899 act by sending out notifications of its restrictions and the informer fees. When corps officials announced the new permit enforcement plan, Reuss told the House of Representatives that the corps was acting vigorously while the Justice Department dragged its feet (515).

The corps also appeared to be trying to enforce the law. Lieutenant Colonel John H. Cousins, Assistant Chief of the Operations Division, Civil Works Directorate, U.S. Army Corps of Engineers, told the House Subcommittee that 516 cases had been referred to the Justice Department under the Refuse Act in fiscal 1970, but the Justice Department did not even inform the corps of its subsequent action. Therefore, Cousins indicated that another dormant provision of the 1899 Act (33 U.S.C. 413) would be reactivated to request that the Justice Department furnish the Secretary of the Army with reports on the disposition of each case. However, the corps also filed litigation only when permit procedures had been administratively exhausted and then in civil instead of criminal proceedings.

As the Corps of Engineers began to show signs of moving to enforce the 1899 Refuse Act during the summer of 1970, President Nixon's Advisory Council on Executive Organization, headed by Roy Ash, former president of Litton Industries, placed a high priority on establishing a Department of Environmental and Natural Resources. The new agency in the executive branch of government would assume control of programs such as pesticide regulations, then under the Department of Agriculture, and water pollution which was under the Department of the Interior. When the EPA was founded in December 1970, personnel and practices were drawn from the predecessor agencies.

While the reorganization was in progress, the corps' request for a $4 million budget to hire personnel and implement the permit program was delayed. Although both the Corps of Engineers and the FWQA had the authority to request that the U.S. Attorney General institute injunctive lawsuits, a period of coordination began instead. After Reuss complained that members of the Environmental Council had been quite vague about the cause of delay in enforcing the permit program, Russell Train replied that "As you know, four agencies (Corps of Engineers, EPA, Department of the Interior, and Department of Justice) are involved; and any program relating to existing discharges requires extensive preparation." On September 17, 1970 Robert E. Jordan, III, Special Assistant for Civil Functions, testified before the House Subcommittee that the Rivers and Harbors Act of 1899 would soon be initiated "in coordination with the Environmental Quality Council and other agencies." But the Justice Department did not even notify U.S. Attorneys that the Corps of Engineers planned to enforce permit requirements under the 1899 Refuse Act (494).

Instead, two days before Christmas 1970, President Nixon announced a new permit plan that Russell Train heralded as "the most important step to improve water quality that this country has taken" (517). When Presidential Executive Order 11574 was published in the *Federal Register* on Christmas Day 1970, new and complex procedures narrowed the scope of the 1899 Refuse Act and offered avenues of evasion, because the program had to be coordinated among several federal agencies. In the first place, the Secretary of the Army was directed to consult with the Administrator of the EPA before authorizing permits for waste discharges into the waterways. The EPA would decide who should get the permits that the corps issued. Henry Reuss notified the President that the corps' authority under a Congressional Act could not legally be delegated to an agency in the executive branch of government (518). Moreover, companies could contend that they were denied due process if one organization authorized the permit and another issued it.

The new Executive Order also specified that the Secretary of the Army should consult with the Secretaries of Interior and Commerce to determine the effects of pollution on fish and wildlife under the Fish and Wildlife Coordination Act of 1958, the National Environmental Policy Act of 1969, and the Federal Water Pollution Control Act of 1970. Before the corps could issue permits under the new program, all the regulations, policies, and procedures of the federal agencies were to be coordinated by the Council on Environmental Quality.

The Presidential Executive Order had also summarized the Fish and

Wildlife Coordination Act so that its scope was significantly narrowed. Although President Nixon had stated that advisements under this act were to be taken when the water was to be impounded, diverted, or the channel deepened, he left out "or the stream or other body of water otherwise controlled or modified *for any purpose whatever by any public or private agency* under Federal permit or license." Reuss pointed out that the Executive Order illegally modified the Congressional Act. However, the Corps of Engineers' new permit plan followed the wording of the Executive Order when it was published in the *Federal Register* a week later (519). Robert Jordan expressed doubt that the Refuse Act was intended to apply to a discharge from a private facility "that did not impound, divert, deepen or similarly modify the body of water," the old navigation issues that had previously been the corps' primary concern. In other words, they still refused to accept responsibility for monitoring environmental contamination.

The permit applications to discharge waste were due by July 1, 1971, and state certification was to be required after April 3, 1973 (519–520). The EPA expected 40,000 applications, but with no provision for enforcement they received 3000. Then the deadline was extended to October 1, and the EPA began an extensive letter campaign to notify companies that they were expected to comply with the new permit program. In late September 1971 the names of 35 companies were forwarded to the Department of Justice to be prosecuted for failure to file permit applications. Two months later, 19,441 applications had been received. Of these 15,292 were from so-called critical industries. Quarles told the Reuss Subcommittee that the companies complied because the EPA has shown a willingness to "go to the mat" and win in the courts (521). But the original list of 40,000 polluters was dwindling fast as companies still avoided compliance by instituting recycling systems, diverting effluent into municipal waste-treatment systems, or closing down. In some cases companies on EPA's list had disappeared, because they filed applications under different names (521). However, Kansas City Water Pollution Control Director Robert Cook injected one positive note. He said the federal crackdown was helping him do his job, because companies that had previously resisted municipal waste treatment were so eager to escape the bother of permit applications that they were clamoring for city assistance. Consequently, they were willing to pay for city treatment of some 6 million gallons of relatively clean cooling water that they had not separated from their more heavily polluted effluent (520).

Despite the slow start, by November 1971, a backlog of 20,000 applications had accumulated, although only ten permits had been issued. Some applications had to be returned because they were incomplete, they had different names, or they even had no names at all. Companies were asked to determine how much and what kind of waste they were discharging, but many did not know. Industrialists had not previously had to collect data on the chemical make-up of their discharges. Instead, waste had generally been measured by the weight of the suspended and dissolved solids or their biological oxygen demand in the waterways (522). This simple measuring system can by no means determine the complicated way the numerous compounds and mixtures affect the ecosystem. Some of them degrade readily and form new combinations, whereas others remain essentially stable for years. And some generate heat, whereas others cause unpleasant tastes, odors, colors, or excessive mineralization.

Moreover, the industrial discharges have increased steadily, although municipal losses to the waterways have declined where federal assistance was contributed to build new waste-treatment facilities. Industry and municipalities discharged approximately equal organic waste loads in 1963, whereas industry discharged 3 times as much in 1970 and consumed half the nation's water requirement. Industry was expected to use 65% by 1989 (522). By 1970 American industries discharged nearly 22 billion gallons of waste daily with only 29% receiving any type of water-purification treatment. These figures were only rough estimates based on limited information about the characteristics and quantity of waste effluent from some plants. Others contributed no information at all.

Monitoring the composition of the changing waste discharges would require new and costly procedures that would constantly have to be updated. In addition, knowing what was there did not insure that the technology was available to treat it. By 1970 Secretary of the Interior Walter Hickel said new chemical pollutants were being added to the nation's waterways at the rate of 55 each year. More than 5000 separate processes had been listed by the Manufacturing Chemists Association as producing waterborne wastes in 1965. Nonetheless, in 1969 the FWQA listed only 51 pollutants that were being introduced into the nation's waterways. After the mercury scare and the subsequent pollution legislation, the EPA was finally forced to publish its initial list of toxic substances at the direction of a federal judge after a lawsuit had been instituted. In the wake of the Minamata disaster and subsequent lawsuits, a Japanese judge ruled that companies that could not clean up their wastes should not operate. In the United States, however, federal agencies ignored the waste discharges for years and then moved very slowly to find out what was being discharged and to form policies not to halt the discharges but to monitor them.

Representative Reuss and Senator Hart held joint subcommittee hearings in February 1971 to consider amendments to the 1899 Refuse Act to close loopholes such as that it only applied to navigable interstate waters. New legislation was needed to monitor water quality in intrastate waterways and ocean outfalls. Consequently, a new round of amendments was proposed to compensate for deficiencies in the Federal Water Quality Control Act and the 1899 Refuse Act. One debate centered on whether the new federal permits would immunize a discharger from prosecution for previous violations. Then companies would only be liable for enforcement of permit conditions in the future, and penalties would have to be stiff enough to insure waste control.

While the new amendments were being debated, the Corps of Engineers received 20,000 applications and issued 25 permits under the old 1899 regulations before the permit-issuing process was halted by a court injunction in December 1971. As a result of the Kalur case, the Army Corps of Engineers also was required to amend its regulations to demand the environmental impact statements specified under the National Environmental Policy Act of 1969. And no permits could be issued for discharges into nonnavigable tributaries of navigable waters. Like Catch 22, in the PICCO case the Third Circuit Court of Appeals ruled in May 1972 that no company could be held criminally responsible for discharges under the Refuse Act until a permit system was in operation (523). With the old permit program stymied, debate over amendments to the Federal Water Quality Control Act generally found the government and industry on one side and academicians and environmental groups on the other as they laid the groundwork for new waste discharge monitoring regulations.

The Justice Department and the National Association of Manufacturers (NAM) favored letting the states monitor pollution. Administration witnesses recommended that Congress integrate the Refuse Act of 1899 into the Federal Water Pollution Control Act, transfer the permit authority to the EPA, and delegate the permit-issuing authority to the states. NAM wanted the EPA's authority under the Refuse Act limited to anchorage and navigation rights. The EPA wanted the states to control the permit program, subject to federal review. However, a November 16, 1970 report from the Office of Enforcement and Standards Compliance said most states already had agencies with the authority to enforce their water quality implementation plans. They just had not done so. Even by February 1971, eight states had no standards for intrastate waters, although a 1966 federal statute had required that they be submitted, and half the standards in the other states were inadequate. Even with federal supervision required, 34 states had unilaterally and illegally extended their pollution abatement deadlines without the EPA's consent. Moreover, some states submitted incomplete information on interstate waters to the federal agency. Where industries dominated the state agencies, officials were reluctant to process applications for state certification of pollution-abatement permit programs under Section 21(b) of the Federal Water Pollution Control Act. Reuss charged that having the states control the permit program "borders on the ridiculous," because they were improperly equipped. Furthermore, 50 different programs would be an enforcement nightmare that would encourage industries to shop around and locate in a less progressive state. In other words, water pollution control would return to square one.

The states' regressive attitude toward pollution abatement had long been influenced by the desire to attract industry to improve the tax base. In the Hart Subcommittee Hearings, Michigan State Senator John Bowman had described the "fierce and frantic competition" among local and state governments to attract industry and the leverage this had given industries to negate pollution abatement.

> Factories have said, and industry has said, if you make us do this, we are going to move. If you make us, by law, put smoke abatement equipment on our chimneys and do certain things that are going to cost us millions of dollars, we are going to move to the State of Ohio. We are going to move to the State of Wisconsin. And those states are saying to industry, you come here and we won't make you do what Michigan may make you do or vice versa (524).

After industries located in a state, they were often well represented on state air and water control boards. In 1970 the *New York Times* conducted a survey that revealed 35 states dotted with industrial, agricultural, municipal, and county representatives whose organizations or spheres of activity were related to pollution. The *Times* reported that "the roster of big corporations with employees on such boards reads like an abbreviated bluebook of American industry, particularly the most pollution troubled segments. . . ."

The state boards were often composed of statutory part-time citizen panels of gubernatorial appointees and state officials who set policies and standards for pollution abatement and then oversaw their enforcement. These state agencies also deal with the federal government (525), and conflicts of interest are frequent. For instance, a hearing on stream pollution by a brewery in Colorado was presided over by the pollution control director of the brewery. And in Los Angeles a member of the committee for harbor pollution was an executive of an oil company that was a chief source of waste discharges. The steel industry, a major polluter of both air and water, is most heavily represented on such boards. The presence of company officials on state boards is often defended, because they are familiar with the pollution problems and rate special consideration because of their civic importance. Average citizens must have very little civic importance, because the general public is seldom represented. Nonetheless, such boards have taken some steps toward pollution abatement, and seven state boards do not permit members with business or professional conflicts. Eight more manage their pollution monitoring with full-time state agencies.

Federal agencies may be less subject to direct industrial manipulation than local and state pollution boards. However, Watergate and the related trials of industry officials for illegal campaign contributions have made the public more aware of the relationship of industry and government at the federal level. Ralph Nader has also warned of the international complications when abatement is attempted. Lax pollution laws in Quebec and Brazil attracted companies like Union Carbide to move plants to these countries instead of complying with pollution control laws in Ohio and Virginia (526). This tactic can also be reversed. In May 1970, when Secretary Hickel warned the German Badische Analin und Soda Fabrik (BASF) Corporation that it might not be able to build a proposed plastics plant on South Carolina estuarine waters in Beaufort County, Manager Designate Dr. Herbert Ende was quoted as saying their $2 billion organization could "flush all of you down the drain. We can wipe you out" (146). Since the European Common Market members signed a pilot anti-dumping pact to prevent further marine pollution, the United States is likely to attract more new business from polluters who do not want to conform to the rules back home.

Whereas the degradation of American waterways has reflected previous state policies that were susceptible to industry pressures, those most visibly represented on pollution control boards also often do the most advertising to show concern for the environment. According to Environmental Action testimony before PL 92-500, the 1972 Water Quality Act, was passed, manufacturing corporations spent over 20 times as much on advertising in 1969 as they did on water pollution cleanup (527).

The New York-based Council on Economic Priorities established that the five industries responsible for over half of such advertisements also were the ones with the worst pollution records. In 1970 the $6 million spent on environmental advertising in all issues of *Time, Business Week,* and *Newsweek* was divided into $3.3 million from the electrical utilities, iron and steel, petroleum, paper, and chemical industries. The paper industry was first with 47 of the 289 pages of advertising. Six of the seven paper companies had been categorized in an earlier council study as having "distinctly unimpressive environmental records" (528). Similarly, in May 1970 the Dow Chemical Company at Sarnia, Ontario, placed a full-page advertisement in the *Detroit Free Press* to explain that it had not known the toxic effects of mercury on fish, that Dow was only one of many companies discharging the element, and that the company was searching for answers because of its great concern for the environment (122). In court, company officials said it would take an act of war with Canada to force compliance with the state of Ohio's request that they remove the pollutants from Lake Erie. The Canadian company contended that it could not be forced to comply, although its central headquarters is on the American side of the St. Clair River in Midland, Michigan.

After 1972

Under the Water Quality Act of 1972, the EPA was delegated the primary responsibility for determining the current water condition, setting standards, issuing permits for discharges, monitoring intentional and accidental discharges, and authorizing research to obtain future abatement. The U.S. Army Corps of Engineers would monitor navigation practices on waterways, their traditional role. The 1899 Refuse Act then was no longer a separate enforcement vehicle. Although simpler and perhaps more appropriate for some of the horse-and-buggy waste discharge procedures still widely employed, more complex technology to monitor discharges and obtain abatement required the formulation of new policies and procedures. When the 1972 Act was passed, a moratorium on enforcement was declared until December 31, 1974 to give the EPA time to set standards and issue permits for waste discharges. The first permits would be written on the basis of existing effluent guidelines and the best technical judgment of ambient conditions (529). They would be in effect for 5 years during which more precise data and plans for abatement would be compiled. As the initial standards were promulgated, the EPA had to thread through cries of too little too late and too much too soon as environmentalists and industrialists squared off with opposite denouncements of the new formulas.

The 1972 Water Quality Act set goals of "the best practicable" waste discharge technology by July 1, 1977 and the "best available" technology by July 1, 1983. To "restore and maintain the chemical, physical, and biological integrity of the Nation's Waters," all discharges of waste effluent into the waterways were to be halted by 1985. Although some flexibility was indicated in the original law, meeting the deadlines appears increasingly unlikely. EPA Deputy Administrator John Quarles acknowledged on January 17, 1974 that a "cold, sober appraisal leads me to conclude, at this stage of the game, that it isn't going to happen the way many of us hoped it would . . ." (530).

The task is monumental to begin with, and the costs are astronomical. The 1972 Water Quality Act initially scheduled $35 billion to meet the abatement goals with $18 billion authorized for 1973–1975. President Nixon vetoed the bill in part because of the cost. After the veto was overridden, he refused to allocate all the funds. Since the federal government was to pay 75% of the municipal waste-treatment plant construction costs, New York City and Detroit sued to make the EPA release the funds. Then a Federal District Court judge ruled that flexibility was needed in spending the money, but all of it should be allocated. In January 1974 the EPA estimated that an additional $543 million in federal funds would be needed to develop the technology to achieve the abatement goals as well as from 5 to 7 additional years for their implementation. The General Accounting Office (GAO) also concluded that the EPA would need more time and money to complete the initially disordered research. Moreover, GAO estimated that bringing sewage treatment plants to the required secondary waste treatment scheduled for July 1, 1977 would take 24 years at the current rate of funding (530). Industry costs just to process the applications for permits to continue dumping wastes into the waterways were estimated at $40 million. Costs for implementing a new self-monitoring program and developing abatement procedures would vary widely, but estimates often run into the billions of dollars for individual industries, even though the EPA allocates government funds to conduct research into abatement techniques. Thus environmental deadlines and goals are likely to be softened if Congress reviews the law as expected (530).

By 1973 the EPA had a work force of 4000 persons, some inherited from the FWQA along with data on the nation's water quality collected since 1962 and some state water quality standards established under the 1965 Water Quality Act (487). As the first step toward meeting the new water quality goals, EPA divided the waterways into two categories. Class A water would be protected for total body contact recreation, and Class B water would allow fish to survive and propagate but would not be suitable for swimming. Although acceptable water quality standards were being violated in one-third of the stream miles in the United States in 1973, most of the waste discharges were concentrated in a few areas. The United States has 2668 major industrial dischargers located on 267 water basins. The water quality was below acceptable standards in 89 of them where 65% of the nation's population and 60% or 1590 of the major industrial dischargers are located. These basins average 18 industries each, whereas the other 178 basins have 30% of the population and 1078 dischargers. This 40% of the total major discharges averaged seven per basin (531).

Funds for sewage-treatment plant construction were sometimes authorized in conjunction with requests that the states formulate plans for basin modelling to determine appropriate waste discharges. This met with some resistance in states where basin modelling was considered a costly matter of low priority. Although they had received up to 55% of municipal wastewater-treatment-facility construction costs since the original Federal Water Quality Control Act was passed in 1948 (483), many municipalities were still struggling to get primary treatment facilities, much less secondary waste treatment by July 1, 1977. The 1972 act also required that the best practicable methods of sewage treatment were to be provided after June 30, 1974. Plants built before that date had until June 30, 1978 to comply with the required secondary treatment (532).

The EPA also wanted model ecosystems as a basis for pollution standards based on individual conditions in the waterways and the technological "state of the art" for each company. A standard based on the individual plant, waterway, and abatement technology was more popular with industry than a single standard for all plants regardless of their obsolescence or potential for abatement. For one thing, $2/3$ of the 300,000 factories that use water are located in the coastal states where the waste can be poured into the oceans. And 10,000 establishments consume 90% of the water. Thus 2% of the industries generate 97% of the total liquid waste. Over half comes from four groups of industrial polluters: the organic chemical, primary metal, paper, and petroleum industries. If all industries had to meet the same standards, the impact for some could be an economic disaster. Therefore, instead of a no-waste discharge standard or individual standards for each company, several compromises have been suggested.

Abatement standards could be set on the so-called "safe pipe" system under which hazardous substances would be limited either arbitrarily or according to the ambient water quality. Then much more waste could be dumped in the ocean than in a small stream, and the quantity of the waste discharge allowed might depend on its toxicity. As mercury would be restricted because it can accumulate in fish, other substances would be limited if they were carcinogenic, mutagenic, or teratogenic. Setting the standard assumed knowledge of these effects, but this was not the case.

After the EPA published its original list of 22 toxic substances, the next step was to propose standards for waste discharges, hold hearings, and then publish the revised standards. Then these often were challenged as too lax by environmentalists who wanted to clean up the waterways and too stringent by the industries that had to pay for some of the cost to develop the technology to implement them. Nonetheless, if mercury serves as the prototype, the standards appear to be quite lenient. When the EPA posted the proposed Toxic Pollutant Effluent Standards in the *Federal Register* on December 27, 1973, the allowable discharges were based on the size of the watercourse. A maximum of 1.62 lb could be dumped per day into rivers, 1.35 lb into lakes, 27.0 lb into estuaries, and 32.4 lb into coastal waters (533).

Chloralkali companies previously taken to court for mercury discharges had agreed to limit their losses to 8 oz per day. If this were technologically feasible, the new standards would not motivate any more abatement, whereas obtaining a permit could prevent prosecution for previous losses. The EPA announced that although almost all companies in the country could technically be prosecuted for previous illegal discharges, lawsuits would not be initiated if they applied for permits (i.e., forgive and forget).

Besides formulating standards for industry, the EPA is charged with monitoring waste discharges from municipalities, agriculture, and commercial enterprises. Their responsibility also includes monitoring accidental spills, ocean dumping, and ground water quality. In 1971 more than 1500 oil spills were counted in the Great Lakes, and 700 oil spills occurred annually in marine waters or contiguous zones of the United States that include salt-water harbors, rivers, and bays.

Ocean dumping includes about 80% dredging spoils, 10% industrial waste, and another 10% smaller discharges such as sewage sludge, construction and demolition debris, solid waste, explosives, chemical munitions, and radioactive waste (531). These can at least be monitored, whereas groundwater contamination is more difficult to control. As industries dispose of more toxic wastes on land or in wells, the risk of groundwater contamination increases. Groundwater supplies about ⅓ of the drinking water for the nation's 100 largest cities and 95% of its rural population as well as over half the water used for livestock and irrigation. Contamination could make groundwaters unfit for use much longer than surface waters with more potential for self-cleansing (531).

With so many varied responsibilities, the EPA is maintaining control of monitoring discharges in ocean outfalls and on federal property, but the states are being encouraged to assume control of the permit program under federal auspices. The states again have primary responsibility for supervision. As some of them were soon authorized to issue the new permits, the EPA restricted its role to offering guidance, providing financial and technical assistance, and reviewing accomplishments. The states also continue to supervise municipal construction of sewage treatment plants and have the primary role in enforcing water quality programs, just like always. However, now the EPA also has the authority to enter industries and collect samples for monitoring the effluent. This pollution discharge information can be made public, but the EPA also indicated that industry trade secrets would be preserved. At the same time, the agency encouraged shared responsibility with the states on enforcement. Unfortunately, areas with more pollution problems that require stringent cleanup standards are still apt to have industry-dominated pollution control agencies. They need federal assistance so enforcement also can be more effective. At least the penalties are greater under the 1972 amendments than under the 1899 Refuse Act. Fines were raised to a minimum of $2500 and a maximum of $25,000 per day and a year in jail. Citizens can institute suits and receive court costs in judgement, but their participation is apt to be limited as they no longer receive awards for giving evidence when the government prosecutes. However, public participation is encouraged "in developing, revising and enforcing all regulations, standards, effluent limits, plans and programs. . . ."

With states again the primary agents for pollution abatement, the potential for enforcement may have been greater under the 1899 Refuse Act. Certainly, the EPA has been overloaded with work. However, the agency did institute uniform procedures for the states to follow in filing water quality information with the federal agency. The state governors were requested to identify one person to report on the status of water quality implementation plans. Nonetheless, in 23 states interim implementation dates still had to be set as permits were issued to begin enforcement in 1975 (534). This was not often expedited, as EPA also showed a willingness to delay procedures somewhat to involve the states even if their program or application and enforcement had not yet been approved. However, the basic permit standards were set by the federal agency, and by December 1973, 27,000 applications had been received, and 1700 permits had been issued. Municipalities were expected to require 35,000 permits (535), and ocean dumpers would receive another thousand (534). Thus the sheer number of permits to be issued makes the task enormous, and even the modest abatement goals initially based on estimated losses have been mitigated. Although the EPA showed a commendable concern for protecting jobs and lowering costs, this can also be interpreted as a lack of resolve for enforcement.

— —(4)

Exceptions to classifications lower than that for recreation and fish and wildlife should be justified. If, based upon natural conditions or upon defensible socio-economic analyses, the desired uses are not possible, exceptions can be made. In these cases, the exceptions should be on specific criteria basis. If, for example, the State adequately demonstrates that natural conditions or man-made conditions that cannot be realistically controlled preclude reaching the desired water quality criterion for turbidity, then only this criterion should be excepted [sic] for the body of water in question (536).

Given the lack of information and the size of the task, enlisting state support to implement the water-quality-control permit program seems the only hope for success despite the economic and political complications raised by joint state and federal responsibility as well as the long tradition of industrial influence on some state pollution control boards. Even in Japan where public resolve became stronger after people died of water pollution-related diseases and new epidemics are still being discovered, it has been difficult to convince government officials as well as industrialists that pollution abatement is worth the cost. In the United States where no major human poisoning epidemics have yet occurred, it is even more difficult to convince people that preventive measures are worth the cost despite such warning signs as increasing numbers of fish kills from acute toxicity and the more gradual increase in mercury concentrations among other creatures. Furthermore, overcoming the long-established precedent for negating or circumventing pollution-abatement efforts will be difficult.

Not enforcing the 1899 Rivers and Harbors Act for 70 years and industrial domination of state water quality control boards has kept manufacturers from having to include pollution abatement in the cost of operation (524–525). Just getting the companies to apply for permits required court action. Now the permit plan is essentially under state control in many instances, whereas the EPA sets general standards and dispenses funds to conduct research into abatement techniques. Companies are being forgiven for their previous discharges without penalties and are being subsidized in efforts to find abatement techniques while enforcement of the new standards is delayed. Thus revising the balance between corporate profit expectations and the need for environmental quality will require more public support. Otherwise, it appears that a great deal of money can be spent as many laws have been passed, agencies set up, and regulations promulgated while the quality of the environment continues to deteriorate.

20

A DEVELOPING PATTERN

ANATOMY OF A POLLUTION PROBLEM

When criteria were being formulated to protect workers from occupational exposure to inorganic mercury in 1973 and steps were being taken to implement the water quality amendments that had been passed in 1972, 4 million lb of the element were still being consumed annually by American industry, agriculture, and medicine. At that time approximately 150,000 people were routinely exposed to mercury in industry. By April 1975 approximately 500,000 lb of elemental mercury were still being added to seed dressings as well as an unspecified quantity to nonmarine paints and other preservatives. In 1974 all American industries consumed 60,070 flasks of mercury, 11% more than in 1973. The demand for mercury in catalysts increased significantly, although the chloralkali industry was not recovering as fast because of the current business recession. Demand had declined markedly in the manufacture of electrical and industrial control instruments where substitutions could eliminate the liquid element. The price had also dropped dramatically. It averaged $117.23 in New York at the end of 1975, and the marginal mines had again closed.

Because many of the applications are not easily replaced in the contemporary society, the demand for mercury is likely to continue to rise. In 1976 the EPA rescinded the ban on mercury in paint before it could actually be implemented, because effective substitutes were not available, and levels in other industries were also rising (see Table 2). Thus the

Year	Chloralkali	Electrical	Paint	Total*
1965	8,753	14,764	7,534	76,454
1966	11,541	13,643	7,762	72,033
1967	14,306	13,823	6,151	69,517
1968	17,453	19,439	8,219	75,422
1969	20,720	18,650	9,486	79,104
1970	15,011	15,952	10,149	61,506
1971	12,262	16,938	8,191	52,475
1972	11,519	15,553	8,190	52,907
1973	13,070	18,000	7,571	54,283
1974	16,816	18,888	6,925	60,070
1975	15,222	16,971	6,928	50,838

*Total Consumption includes: Agriculture, catalysts, dental preparations, general laboratory use, industrial and control instruments, pharmaceuticals, and other uses that are not specifically identified. Bureau of Mines statistics.

Table 2. Major American Mercury Uses and Total Industrial Consumption in flasks, 1965–1975

natural supply continues to dwindle as the environment is further degraded. The partial pollution-abatement standards set since 1970 have permitted industries to evade the high cost of waste-reclamation equipment in many cases despite the millions of dollars expended by the federal government to establish monitoring procedures and to pay for research into new abatement methods. Unless total waste-recycling systems are implemented to clean up the waterways and save the valuable components in the waters, a scarcity of resources can be engendered along with a waste-disposal problem. At the same time, the pattern of evasion and compromise between industry and government that Jun Ui noted in Japan can be repeated in the United States.

Despite the enormous financial outlay for researching and monitoring mercury pollution, contemporary economic patterns tend to be favored over enforcing environmental standards. In Japan, after the first poisoning epidemic from mercury-contaminated fish forced a new social awareness of this environmental hazard, in several subsequent epidemics Professor Ui described the typical approach to a pollution problem. It generally follows four basic steps: outbreak, identification, refutation, and solution. The last step is often evaded rather than completed.

The initial outbreak of mercury poisoning at Minamata, Japan, was kept secret as long as it appeared only to affect individuals and families at the lower end of the social scale. However, when an epidemic appeared to be in progress and several people died, the disaster was widely publicized and a public outcry was raised to find the cause. Similarly, in the United States the television picture of Amos Huckleby, a permanently crippled child victim of mercury poisoning returning home from the hospital, was enough to have alkylmercury fungicides immediately banned from seed dressings.

Because a veil of secrecy and ignorance often surrounds an outbreak of environmental mercury poisoning or the contamination itself if the problem is identified before people are actually suffering, the first public warning may be a press release in which government officials deny that the problem exists for one reason or another. At this stage a list of defenses generally is raised to prevent the collection of evidence. Officials of government and industry often contend that the pollutant occurs naturally in the environment and so is relatively harmless or may actually be beneficial. Although no supporting evidence has been advanced, some American authorities have regularly theorized that mercury may be a necessary micronutrient in the human diet. The toxic impact of a contaminant is also often distorted quantitatively by not clarifying how much or what form causes the damage to human health. Thus miners have swallowed several ounces of elemental mercury as a home remedy to clear their intestines, whereas minute quantities of methylmercury permanently damage the human brain and central nervous system. An additional quantitative defense is that even "pure" air and water can be toxic in large enough quantities. All these defenses circumvent the real issue.

Moreover, although mercury pollution of the environment has been identified at an earlier stage in each country, this may have made the problem easier to avoid than solve. Whereas a public outcry was raised when an epidemic was in progress and people died or were crippled at Minamata, in Sweden the deaths of seed-eating birds warned of a long-term problem of high mercury levels in food for human consumption rather than immediate physical jeopardy. And the problem seemed even more remote in Canada where an exchange of information with Swedish scientists motivated studies of mercury levels in apparently healthy wild fowl and fish. The Canadians, in turn, warned the Americans. Less short-term hazard generates less public concern and relegates the problem to obscurity more quickly after the initial publicity declines at the second stage of the pattern: identification. For one thing, people are apt to have faith that discovery of the cause automatically insures finding a cure as well. Thus a false sense of security is easy to foster at the third stage: refutation of whatever evidence has surfaced.

Once the hazard is publicized by the private citizens who are directly affected, state and federal officials express concern and a desire for immediate action. "Yes, we have a problem and are trying to solve it." This is the stage wherein restrictions may be hastily implemented to prevent further environmental degradation and protect public health, but instant verbal panaceas cannot eliminate contamination that has accumulated over several decades. Moreover, the strict new policies are often compromised and relaxed, sometimes before they are even totally implemented if the economic impact is quickly felt. For instance, in the first American mercury crisis at Lake St. Clair, the fishing ban was soon changed to a Catch-and-Release policy because of the economic losses to the local tourist and sport-fishing industries. Subsequently, other states were not as aggressive in the search for environmental contaminants or in publicizing them because of the likely negative impact. Instead, they set more liberal standards and presented the rationale that the problem was not serious or the standards were relaxed later because the mercury levels were reported to be declining in fish. The new evidence that methylmercury concentrates in the brain for a very long time is generally disregarded. Otherwise, the standards would become more strict with lower acceptable levels of mercury in food.

Once the problem is acknowledged to exist, the public is usually reassured that it is not as bad as had been indicated. Then the media are often accused of having overreacted. Opposing researchers also confuse the issue because the evidence is not conclusive. Professor Ui contends that university researchers often appear to be objective third parties who happen to present evidence that helps the polluters. But in reality, scholars also have a vested interest, because they receive consulting fees and research grants. For whatever reason, the scientists' carefully qualified statements may confuse the public further on how serious the problem is.

Workers in polluting industries may also be both victims and victimizers, although Ui draws some distinctions between the two categories. The victimizers benefit in some way from continuing the pollution, and they have the power to cloud and confuse the issue. On the other hand, the victims want to clarify the cause and effect. With little power they must dedicate their lives and resources to achieving reform, a cause that is only a small issue in the lives of the more powerful polluters. At Minamata the victims, the poor Japanese fishermen, were essentially engaged in a life-and-death struggle, whereas the victimizers, the giant Chisso Company, had support from both local and federal agencies. City assemblymen often represented either Chisso's management or one of the several unions. Similarly, in the United States local governments compete to build their tax base by attracting more industry and may ignore their own pollution laws to achieve their goal. This undercuts citizens who institute lawsuits to prevent the further degradation of their health or property. The states also may be relaxing their monitoring efforts to attract or retain industry at the same time that federal agencies are trying to impose more restrictions. Or the opposite may be true, so industries can take advantage of conflicting standards to continue polluting the environment.

With most of the power on one side of the struggle, the victimization is likely to reach ominous proportions before it is publicly acknowledged. At a conference in Japan, visiting social scientists talked blithely about environmental problems at the hotel, but they were overwhelmed and helpless when confronted with the awesome reality. Even when the victims have support, such as when Japanese students championed their cause at Minamata, restitution usually takes the form of monetary compensation, often minimal, that does not restore the victims' health or means of earning a living. Nor does it deter further pollution if no industrialists are prosecuted or reforms instituted.

Instead, what generally follows is a proliferation of laws that spawn agencies and concentrate power in the hands of the central government, ostensibly because local and state agencies lack adequate technology to monitor the problem. But environmental conditions are apt to remain the same or even grow worse. In Japan the so-called "Pollution Diet" passed 14 laws by the end of 1970, but direct restrictions on polluters were minimal because the government did not want to deter industrial progress. Similarly, in the United States the new battery of federal laws vested much of the authority to monitor pollution in the EPA. Their task is so large that the individual states have been requested to monitor their own programs under EPA supervision, but this is often ineffective. At one point 34 states had illegally extended compliance deadlines without the EPA's approval.

In addition to political compromises, industrial research to decrease pollution has been markedly but not surprisingly slower than the profit-making phases. Japanese industrialists often borrowed manufacturing procedures from other technologically advanced societies, but Ui charges that they seldom adopted even the limited pollution-abatement technology that these countries had developed. Instead, such procedures were only slowly implemented under extreme pressures. American pollution-abatement laws have also been passed in faith that the technology could be devised to meet their requirements. Then the deadlines have often been extended, because appropriate procedures were not yet devised or proved to be damaging in other ways, as with the automobile emissions-control devices.

To explain why the contamination was allowed to continue for years, often in the face of state and federal regulations that were not enforced, a number of reasons are commonly cited. One is that the responsibility rests with someone else: another department, an outside state or federal agency, or some irrational so-and-so upstairs. Thus the bureaucrat says "It's not my job," and the scientist says "I just compile the data." Industry pleads ignorance of possible harm from contamination and a harried willingness to comply with conflicting or excessively severe government standards. The respective agencies, state or federal, generally contend that improvement in the environment depends on meeting certain requirements such as more experts, refined analytical techniques, equipment, money, time, cooperation, and authority. Moreover, current laws may be nonexistent, inadequate, or conflicting, and new laws do not necessarily improve the situation because they set new precedents, raise the possibility of being prosecuted for something done before the law was passed, and open the way for a multitude of conflicting interpretations.

In addition, restrictions on pollution discharges can also be delayed while adequate data are collected. In the future this problem may be reversed. If all the proposed data are kept, sheer volume may also engender new conflicts in opinion over the best laboratory procedures and the interpretation of results such as the synergistic and antagonistic reactions among pollutants. As new laws and implementation plans have been devised in the United States, often the accumulation from previous pollution has been forgiven with no requirement to clean up the waste. Therefore, the degraded waterways are left to clean themselves, a natural, cheap process that may require decades. Meanwhile, new discharge standards are tentatively based on current estimated losses from industries.

The fourth stage of the pollution pattern, the solution, is generally cloaked in long-term compromises for gradual pollution abatement combined with current research into better pollution control methods. In the short run the water and air are becoming more degraded, whereas land disposal exposes the soil and underground waters to contamination as well.

As the pattern is repeated over and over again with mercury and other elements or compounds in vastly more complex waste discharges, similar results can be expected as other countries follow the prototype of Japan. As the environment becomes more degraded, the cycle is repeated in subsequent pollution-related epidemics: outbreak, identification, refutation, and the promise of a solution that may be too distant to have any meaning. Nonetheless, Jun Ui, rightly called the Ralph Nader of Japan, contends that measures can be taken quickly and effectively to halt the danger of pollution to human beings and their environment.

The first step is to recognize that the problems now affecting the poor fishermen in Japan, the farmers in Iraq, and the Indians in Canada are harbingers of what can happen to everyone as our environment becomes more contaminated. Monetary compensation for the victims cannot help the damaged bodies of the Japanese or return the Indians' traditional way of life. But compensation is a first step to force the companies to accept responsibility for their waste and to make the government implement a genuine pollution-abatement policy that will reverse the downward spiral in environmental quality. Ui contends that the only real way to end contamination is for the public to exert sufficient pressure to place responsibility directly on the polluting industries to end discharges, clean up previous losses, and pay for the damage. The citizens of the world must resolve that the bodies and minds crippled at Minamata shall not have been in vain. To enforce this resolve, the public must not expect those who allowed the contamination to accumulate in the first place to reverse this trend unless we all apply great and continuous pressure to reinforce their determination.

AFTERWORD

by Jun Ui

The future of our environment is not unconditionally bright at present. In fact, there is a clear tendency toward more serious environmental problems during our children's lives, because the environment does not recover easily once it is polluted. This fact was clearly observed in the history of mercury pollution. Although our generation has a strong responsibility to prepare a clean environment for our children, pollution from industrial countries is gradually spreading on a global scale. This creates new problems of international responsibility within the value system of Western, industrialized cultures.

Among many irreversible instances of pollution, mercury is perhaps the most well known in the history of mankind because of the abundant experiences from many recorded disasters. Recent, typical cases are Minamata disease caused by the discharge of industrial waste and the Iraqi poisoning caused by the misuse of agricultural chemicals. Such disasters with a large human toll have been repeated often in recent years, and this repetition clearly shows the importance and severity of the mercury problem. It is necessary to investigate the process of mercury contamination to prevent other large-scale chemical pollution in the future.

Minamata disease, the first widely known mercury pollution by industrial wastewater, was discovered in 1956, but its simple cause-and-effect relationship was not officially recognized by the government until 1968. One may think this delay very strange and ask why it took so long to determine such simple facts unless the real historical process involved in solving the problem is explained. The research group in the medical school of Kumamoto University met with every kind of difficulty in their investigation after the disease was discovered in 1956. First, Minamata Bay, where the polluted fish were discovered, was heavily contaminated by so many kinds of industrial pollutants from the chemical factory Chisso Company that it was nearly impossible to find out what the principal pollutant was. Second, the factory refused to disclose the necessary information on its chemical processes to the research group in the name of protecting trade secrets. Even an inspection of the factory was refused. Third, the government did not give adequate financial support to the research group and many times positively hindered the research by political pressure. Such hindrance by false statements of objections and contradictions was presented not only by the factory but also by other researchers in medical science, chemistry, and so on. The majority of the Japanese scientists had strong ties with the powerful industries and the government. Therefore, they acted, intentionally or unintentionally, as the agents of these strong powers. Such circumstances have not been unusual in the history of Japanese science and technology, but Minamata disease was really the most obvious case of such social, criminal action by scientists to support the incumbent power and to sell the truth for money. As a result of this hindrance, the truth about Minamata disease was completely left out of the public news after 1960, and very limited research into the cause and the effect on human health was continued in the medical school of Kumamoto University.

The same process was repeated in the history of the second Minamata disease epidemic at Niigata in 1965. The victims of the second epidemic, however, tried to elucidate the true cause of the disease by their own action. They filed a civil case against the polluting company in an effort to clarify the cause-and-effect relationship and to establish the legal responsibility of the polluting company through open arguments in court. As a side effect of this action by the victims, the cause of the first disease was also confirmed. It should be noted that the victims finally solved the problem, and the result was propagated to the general public by the court, although the majority of the work to establish the cause-and-effect relationship was done by scientists. Their results coincided with the instincts of the victims. Frequently, the work of scientists is not properly evaluated in public unless it is supported by social action on the part of the victims and the general public.

During the more than a dozen years' history of Minamata disease, many formal organizations which have as their purpose to prevent or to reduce pollution or to promote the public welfare have exhibited behavior that has been recognized as a common pathological reaction. This was named the "Minamata syndrome" by Professor D. R. Thurston of the State University of New York. Not only the polluting company but also the local and central government and sometimes even the trade unions had a common tendency to evaluate the problem as being less important than the actual situation indicated. The victims' complaints were ignored, mutually contradictory measures were applied, legal control always had hidden loopholes, and the previously accumulated evidence was forgotten. It might be a coincidence, but a strange, parallel relationship was observed between the actual symptoms of Minamata disease and the reactions of these formal organizations. A constriction of the visual field was common among all organizations. Ataxia, a loss of coordination between various parts of the body, was often exhibited in contradictions between the measures taken by various parts of the government. There was also a loss of sensation as the appeal of the victims went unheard, and there was little effort to grasp the situation as a whole. Many organizations also reacted with spastic convulsions when they faced the problem. This was followed by mental retardation and forgetfulness. Thus what Professor Thurston called Minamata syndrome has a symbolic meaning and includes a wide set of social reactions. This syndrome has been observed not only at Minamata but also at Niigata as well as in other countries such as Sweden, Finland, Canada, and Italy. I reached this widely recognized conclusion after we investigated all these countries, to some degree, in an international comparative study of pollution. Therefore, it seems to us that the Minamata syndrome is a rather common social reaction in many countries and may be a serious problem as it affects how pollution is considered.

The narrowness of the visual field was also observed in the medical cognition of Minamata disease. Here it was an inevitable difficulty that reflected conditions in the research process. The situation was also delicate in the argument over the cause-and-effect relationship, because the local society, which was totally dependent on a strong company, reacted critically. This created difficulties in the process of recognizing the victims of poisoning and identifying the complete set of methylmercury poisoning symptoms, although they had been recorded in previous scientific literature and had been applied to diagnose such patients for a long time. In the 1960s the official policy of the research group was to examine the clinical picture in a restrictive way. Most of the medical reports were written under this policy and printed in foreign medical journals. Since 1970 this restrictive cognition of Minamata disease was drastically revised after the polluted area was carefully observed by a few scientists and the victims themselves. Irreversible damage to the peripheral nervous system, high blood pressure, hardening of the blood vessels, liver damage, heart attack, diabetes, and mental retardation were all nonspecific symptoms that were clearly observed among inhabitants with greater frequency in polluted than in control areas. Some of them were also recognized as the result of methylmercury poisoning in animal experiments. This expansion of categories of symptoms inevitably increased the social responsibility of the polluting company and its economic burden of compensation. This brought out a delicate political question. Therefore, some medical experts are still very reluctant to recognize the results of epidemiological studies. In the past nonspecific symptoms were used as a reason to reject the victims' appeal for recognition, so this change in policy means a loss of face among authoritative experts who were commissioned to ascertain official recognition of the disease symptoms. It also contradicts the policy of the government and the company to limit the problem as much as possible. Therefore, the extensive epidemiological study of the polluted area has been intentionally overlooked by the government, whereas previous studies were insufficient in scope. Moreover, on the basis of political considerations, the third Minamata disease epidemic was rejected after it occurred in the area of Ōmuta and Tokuyama. Politics was also the reason why the health hazard of mercury pollution was underestimated in Sweden and Italy. However, a series of attempts have been made to overcome this difficulty. One such was by Professor Masazumi Harada (537). Unfortunately, these efforts are not well known in other parts of the world yet. Although many reports were published and much was said about the clinical picture of Minamata disease, the problem is still not solved.

In contrast to the powerlessness of formal organizations, the human effort is really impressive where individuals had to face the difficulty and tried to solve the problem. Dr. Hosokawa was the discoverer of the disease. As the most relied upon clinical practitioner in the factory hospital, he tried to find the truth in spite of his delicate position, even against the wish of the factory management. Dr. Hosokawa finally found the true cause-and-effect relationship in 1962, but it was practically impossible to publish his results. They were finally published shortly before his death from lung cancer in 1970. His work taught us the importance of the scientist's humanity and responsibility. In addition, when a blind poet, Mrs. Michiko Ishimure, sang about the life and misery of the Minamata victims in a fantastic, beautiful style, she played a large part in spreading recognition of the problem to all of Japan. However, the greatest effort was the victims' movement to elucidate the truth and responsibility of the company and the government. When the depressed victims finally stood up from the bottom of the society, their movement was by no means organized, but it was supported at the grass roots by many people in Japan, and its steadiness gave a strong stimulus to many other antipollution movements. That these most depressed poor people could act so forcefully is our reason for optimism about the future of mankind.

The basic way to stop pollution is not to establish a formal organization or promote scientific investigations, but it is really the human effort of individuals who have the courage to see the truth and confront the difficulties, although they stand at the edge of disaster. Preparing this book is surely one such sincere human effort, and I hope it will help to change the miserable situation of the Canadian Indians who also suffer from mercury pollution on their remote reserves. Their conditions should be improved as soon as possible.

BIBLIOGRAPHY

1. Steinfeld, J. L., Testimony presented at the hearings before the Sub-committee on Energy, Natural Resources and the Environment of the Committee of Commerce, United States Senate, Ninety-First Congress, 2nd Session on *The Effects of Mercury on Man and the Environment*, Part 3, Serial No. 91-73, Washington, D.C., U.S. Printing Office, 1970.

2. Hamilton, E., *Mythology*, New York, New American Library, 1953.

3. Farber, E., *The Evolution of Chemistry*, New York, Ronald, 1952.

4. Secundus, C. P., On minimum and the humour that is known as quick-silver, Book 33 of Pliny's *Natural History*, 77 A.D., English version by John Bostock and H. T. Riley, 1955. In E. L. Egenhoff, *Calif. J. Mines Geol.*, **49**:28, 1953.

5. Theophrastus' History of Stones, English version and notes by John Hill, London, 1946. In E. L. Egenhoff, *Calif. J. Mines. Geol.*, **49**:12, 1953.

6. Bennett, E., Almadén-world's greatest mercury mine, *Min. Metall.*, **29**:6, 1948.

7. Stillman, J. M., *The Story of Early Chemistry*, New York, D. Appleton, 1924.

8. Li Ch'iao-P'ing, *The Chemical Arts of Old China*, Easton, American Chemical Society, 1948.

9. D'Itri, F. M., *The Environmental Mercury Problem*, Cleveland, CRC, 1972.

10. Leicester, H. M., *The Historical Background of Chemistry*, New York Wiley, 1961.

11. Erckern, L., Spagyrick Laws. The Second Part, Containing Essays on Metalick Words: Alphabetically Composed as a Dictionary to Lazarus Erckern. Translation by Sir John Pettus, 1683. In E. L. Egenhoff, *Calif. J. Mines. Geol.*, **49**:73, 1953.

12. Alchemistes of Araby: Chapter II in which are Set Forth Discourses Upon the Stone of Philosophers: The Part Argentvive. Translation by

M. Berthelot, 1889. In E. L. Egenhoff, *Calif. J. Mines. Geol.*, **49**:31, 1953.

13. Berthelot, M., *Collection des Anciens Alchimistes Grecs*, Volumes 1 and 2, Paris, G. Steinheil, 1888.

14. Jaffee, B., Theophrastus Paracelsus (1493–1541). In *Crucibles: The Story of Chemistry from Alchemy to Nuclear Fission*, Greenwich, Fawcett, 1961.

15. Gerhardt, P. (Hiebert), *Sarah Binks*, New York, Oxford University, 1947.

16. Kurland, L. T., Faro, S. N., and Siedler, H., Minamata Disease: The outbreak of a neurological disorder in Minamata, Japan, and its relationship to the ingestion of seafood contaminated by mercuric compounds, *World Neurol.*, **1**:370, 1960.

17. Takeuchi, T., The relationship between mercury concentrations in hair and the onset of Minamata Disease. In R. Hartung and B. D. Dinman, Eds., *Environmental Mercury Contamination*, Ann Arbor, Ann Arbor Science, 1972.

18. Doi, R., and Ui, J., The distribution of mercury in fish and its form of occurrence. In P. A. Krenkel, Ed., *Heavy Metals in the Aquatic Environment*, Oxford, Pergamon, 1975.

19. Anon., Another outbreak of Minamata Disease, *Chem. Eng. News*, **51(23)**:7, 1973.

20. Kojima, K., and Fujita, M., Summary of recent studies in Japan on methylmercury poisoning, *Toxicology*, **1**:43, 1973.

21. Nomura, S., Epidemiology of Minamata Disease. In *Minamata Disease*, Study Group of Minamata Disease, Kumamoto University, Japan, 1968.

22. Ui, J., Discussion of K. Irukayama's paper, *Advan. Wat. Pollut. Res.*, **3**:167, 1967.

23. McAlpine, D., and Araki, S., Minamata Disease—an unusual neurological disorder caused by contaminated fish, *Lancet*, **2**:629, 1958.

24. Anon., Prof. bares mercury pollution spreading, *The Japan Times*, November 13, 1972.

25. Takeuchi, T., Biological reactions and pathological changes of human beings and animals under the conditions of organic mercury contamination. In R. Hartung and B. D. Dinman, Eds., *Environmental Mercury Contamination*, Ann Arbor, Ann Arbor Science, 1972.

26. Ganther, H. E., Goudie, C., Sunde, M. L., Kopecky, M. J., Wagner, P., Oh, S.-H., and Hoekstra, W. G., Selenium: Relation to decreased toxicity of methylmercury added to diets containing tuna, *Science*, **175**:1122, 1972.

27. Irukayama, K., Minamata Disease as a public nuisance. In *Minamata Disease*, Study Group of Minamata Disease, Kumamoto University, Japan, 1968.

28. Kitamura, S., Determination on mercury content in bodies of inhabitants, cats, fishes, and shells in Minamata district and in the mud of Minamata Bay. In *Minamata Disease,* Study Group of Minamata Disease, Kumamoto University, Japan, 1968.

29. Ui, J., Mercury pollution of sea and fresh water: Its accumulation into water biomass, *Rev. Int. Oceanogr. Med.,* **22–23**:79, 1971.

30. Ui, J., and Kitamura, S., Mercury in the Adriatic, *Mar. Pollut. Bull.,* **2**:56, 1971.

31. Nuorteva, P., Methyl-mercury in nature's food chains, a current problem, *Nordenskiold Soc. J.,* **29**:6, 1969. Fisheries Research Board of Canada, Translation Series No. 1533, Winnipeg, Manitoba.

32. Hunter, D., Bomford, R. R., and Russell, D. S., Poisoning by methylmercury compounds. *Quart. J. Med.,* **33**:193, 1940.

33. Hunter, D., and Russell, D. S., Focal cerebral and cerebellar atrophy in a human subject due to organic mercury compounds, *J. Neurol. Neurosurg. Psychiat.,* **17**:235, 1954.

34. Irukayama, K., Fujiki, M., Kai, F., and Kondo, T., Studies on the origin of the causative agent of Minamata Disease, II. Comparison of the mercury compound in shellfish from Minamata Bay with mercury compounds experimentally accumulated in control fish, *Kumamoto Med. J.,* **15**:1, 1962.

35. Curley, A., Sedlak, V. A., Girling, E. F., Hawk, R. E., Barthel, W. F., Pierce, P. E., and Likosky, W. H., Organic mercury identified as the cause of poisoning in humans and hogs, *Science,* **172**:65, 1971.

36. Harada, Y., Congenital (or fetal) Minamata Disease. In *Minamata Disease,* Study Group of Minamata Disease, Kumamoto University, Japan, 1968.

37. Murakami, U., Organic mercury problem affecting intrauterine life. In M. A. Klingberg, A. Abramovici, and J. Chemke, Eds., *Advances in Experimental Medicine and Biology,* Volume 27, New York, Plenum, 1972.

38. Takeuchi, T., Research Committee on Minamata Disease: Pathological Clinical and Epidemiological Research about Minamata Disease, 10 Years after, 2nd ed. Translation by Dr. Ryotaro Ishizaki, Interuniversity Consortium for Environmental Studies, Durham, North Carolina, March 1974.

39. Tejning, S., Mercury contents in blood corpuscles and in blood plasma in persons consuming large quantities of commercially sold saltwater fish. Report No. 680529, Department of Occupational Medicine, University Hospital, Lund, Sweden, 1968 (in Swedish).

40. Tejning, S., Mercury levels in blood corpuscles and in plasma in "normal" mothers and their new-born children. Report No. 680220, Department of Occupational Medicine, University Hospital, Lund, Sweden, 1968, (in Swedish).

41. Ramel, C., Methylmercury as a mitosis disturbing agent, *J. Jap. Med. Ass.*, **61**:1972, 1969.

42. Ramel, C., and Magnusson, J., Genetic effects of organic mercury compounds, II. Chromosome segregation in *Drosophila melanogaster, Hereditas,* **61**:231, 1969.

43. Ramel, C., Genetic effects of organic mercury compounds, I. Cytological investigations on *allium* roots, *Hereditas,* **61**:208, 1969.

44. Ramel, C., Genetic effects of organic mercury compounds, *Oikos,* **9 (Suppl)**:35, 1967.

45. Anon., Can Aging be Cured? *Newsweek,* **81(16)**:56, April 16, 1973.

46. Thurston, D. R., Aftermath in Minamata, *The Japan Interpreter,* **9**:27, 1974.

47. Johnels, A. G. and Westermark, T. Mercury contamination of the environment in Sweden. In M. W. Miller and G. G. Berg, Eds., *Chemical Fallout,* Springfield, Thomas, 1969.

48. Larsson, J. E., *Environmental mercury research in Sweden,* Swedish Environmental Protection Board, Stockholm, 1970.

49. Borg, K., Wanntrop, H., Erne, K., and Hanko, E., Alkylmercury poisoning in terrestrial Swedish wildlife, *Viltrevy,* **6**:301, 1969.

50. Berglund, F., Berlin, M., Birke, G., Cederlof, R., von Euler, U., Friberg, L., Holmstedt, B., Jonsson, E., Luning, K. G., Ramel, C., Skerfving, S., Swensson, A., and Tejning, S., Methyl mercury in fish, *Nord. Hyg. Tidskr.,* **4 (Suppl)**:1, 1971.

51. Borg, K., Wanntorp, H., Erne, K., and Hanko, E., Mercury poisoning in Swedish wildlife, *J. Appl. Ecol.,* **3 (Suppl)**:171, 1966.

52. Grolleau, G., and Giban, J., Toxicity of seed dressings to game birds and theoretical risks of poisoning, *J. Appl. Ecol.,* **3 (Suppl)**:199, 1966.

53. Westöö, G., Methylmercury compounds in animal foods. In M. W. Miller and G. G. Berg, Eds., *Chemical Fallout,* Springfield, Thomas, 1969.

54. Tejning, S., Mercury in pheasants (*Phasianus colchicus* L.) deriving from seed grain dressed with methyl and ethyl mercury compounds, *Oikos,* **18**:334, 1967.

55. Löfroth, G., and Duffy, M. E., Birds give warning, *Environment,* **11**:10, 1969.

56. Westöö, G., Mercury in pork, calf, beef, and reindeer muscle and liver, *Var Foeda,* **18**:88, 1966; *Chem. Abstr.,* **73**:44020t, 1966.

57. Westöö, G., Mercury and methylmercury levels in some animal food products, August 1967–October 1969, *Var Foeda,* **21**:137, 1969; *Chem. Abstr.,* **72**:109925t, 1970.

58. Wanntorp, H., Borg., K., Hanko, E., and Erne, K., Mercury residues in wood-pigeons (*Columba p. palumbus* L.) in 1964 and 1966, *Nord. Vet. Med.,* **19**:474, 1967.

59. Borg, K., Erne, K., Hanko, E., and Wanntorp, H., Experimental secondary methylmercury poisoning in the goshawk (*Accipiter g. gentilis* L.), *Environ. Pollut.*, **1**:91, 1970.

60. Mellanby, K., *Pesticides and Pollution*, London, Collins, 1967.

61. Loosmore, R. M., Harding, J. D. J., and Lewis, G., Mercury poisoning in pigs, *Vet. Rec.*, **81**:268, 1967.

62. Gurba, J. B., Use of mercury in Canadian agriculture. In *Mercury in Man's Environment*, Ottawa, Royal Society of Canada, 1971.

63. Hill, W. H., A report on two deaths from exposure to the fumes of a diethyl mercury, *Can. J. Pub. Health*, **34**:158, 1943.

64. Lungren, K. D., and Swensson, A., Occupational poisoning by alkylmercury compounds, *J. Ind. Hyg.*, **31**:190, 1949.

65. Ahlborg, G., and Ahlmark, A., Chemical aspects of poisoning by alkylmercury compounds and risk of exposure, *Nord. Med.*, **41**:503, 1949.

66. Engleson, G., and Herner, T., Alkyl mercury poisoning, *Acta Paediat. Scand.*, **41**:289, 1952.

67. Damluhi, S. F., and Tikriti, S., Mercury poisoning from wheat, *Brit. Med. J.*, **1**:804, 1972.

68. Jalili, M. A., and Abbasi, A. H., Poisoning by ethyl mercury toluene sulphonanilide, *Brit. J. Indust. Med.*, **18**:303, 1961.

69. Al-Kassab, S., and Saigh, N., Mercury and calcium excretion in chronic poisoning with organic mercury compounds, *J. Fac. Med. Baghdad*, **4**:118, 1962.

70. Damluhi, S., Mercurial poisoning with the fungicide Granosan M, *J. Fac. Med. Baghdad*, **4**:83, 1962.

71. Haq, I. U., Agrosan poisoning in man, *Brit. Med. J.*, **1**:1579, 1963.

72. Ordonez, J. V., Carrillo, J. A., Miranda, M., and Gale, J. L., Epidemiological study of an illness in the Guatemala highlands believed to be encephalitis, *Of. Sanit. Panam. Bol.*, **60**:510, 1966.

73. Anon., EPA pesticide "bans" questioned, Report B-133192, U.S. Comptroller General's Office, Washington, D.C., 1973.

74. Nelson, G., Statement of U.S. Senator Gaylord Nelson on H. R. 10729, the Federal Environmental Pesticide Control Act of 1971, and amendments, before the Senate Commerce Subcommittee on the Environment, Senator Philip A. Hart, Chairman, June 15, 1972.

75. Bakir, F., Damluhi, S. F., Amin-Zaki, L., Murtadha, M., Khalidi, A., Al-Rawi, N. Y., Tikriti, S., Dhahir, H. I., Clarkson, T. W., Amith, J. C., and Doherty, R. A., Methylmercury poisoning in Iraq, *Science*, **181**:230, 1973.

76. Dillman, T., and Behr, P., Mercury poisoning in Iraq termed a disaster, *The Lansing State Journal*, Lansing, Michigan, March 25, 1972.

77. Anon., Iraqi poisonous seed might have originated in Canada: Official, *The Toronto Globe and Mail*, March 10, 1972.

78. Clarkson, T. W., Department of Radiation Biology and Biophysics, University of Rochester, Rochester, New York, Personal communication, 1973.

79. Shapiro, R. E., Epidemiology Unit, Bureau of Foods, Department of Health, Education and Welfare, Public Health Service, Food and Drug Administration, Washington, D.C., 20204, Personal communication, 1973.

80. Alpert, D., Likosky, W. H., Fox, M., Pierce, P. E., and Hinman, A. R., Organic mercury poisoning—Alamogordo, New Mexico. Viral Diseases Branch, Epidemiology Program Communication, Public Health Service-HSM-NCDC-Atlanta EPI—70-47-2, March 27, 1970.

81. Blumenthal, R., In 18 months, mercury-poisoned girl almost well, *The New York Times*, June 6, 1971.

82. Roueche, B., Annals of medicine—insufficient evidence, *New Yorker Mag.*, **46**:65, August 22, 1970.

83. Snyder, R. D., Congenital mercury poisoning, *New Eng. J. Med.*, **284**:1014, 1971.

84. Martin, H., The Mad Hatter visits Alice's restaurant, *Today's Health*, **48**:39, 1970.

85. Anon., Status report—mercury seed treatments, *Amer. Seed Trade Assoc. News Letter*, **22(17)**:1, March 6, 1970.

86. Kirkpatrick, R., Michigan Department of Agriculture, Manley Miles Building, Harrison Road, East Lansing, Michigan, Personal communication, 1971.

87. Blojer, H. P., Memorandum, Subject: Study No. S-2129 Cerosan-treated wheat seed poisoning cases. Mexicali, Baja California, Mexico, California, Department of Public Health, July 31, 1969.

88. Ottoboni, F., Memorandum, Subject: Mercury treated grain problem. San Diego, California Department of Public Health, October 21, 1970.

89. Westöö, G., Determination of methylmercury compounds in foodstuffs: I. Methylmercury compounds in fish, identification and determination, *Acta Chem. Scand.*, **20**:2131, 1966.

90. Katz, A., Mercury pollution: The making of an environmental crisis. *CRC Crit. Rev. Environ. Control*, **2**:517, 1972.

91. Löfroth, G., Methylmercury. A review of health hazards and side effects associated with the emission of mercury compounds into natural systems. Ecological Research Committee Bulletin No. 4, Swedish Natural Science Research Council, Stockholm, 1969.

92. Johnels, A. F., Olsson, M., and Westermark, T., Mercury in fish: investigations on mercury levels in Swedish fish, *Var Foeda,* **19**:67, 1967.

93. Hasselrot, T. B., Report on current field investigations concerning the mercury content in fish, bottom sediments, and water, *Rep. Inst. Freshwater Res. Drottningholm*, **48**:102, 1968.

94. Hughes, W. L., A physiochemical rationale for the biological activity of mercury and its compounds, *Ann. NY Acad. Sci.*, **65**:454, 1951.

95. Miettinen, J. K., Tillander, M., Rissanen, K., Miettinen, V., and Okanomo, Y., Distribution and excretion rate of phenyl- and methylmercury nitrate in fish, mussels, molluscs and crayfish. In *Proceedings of the 9th Japanese Conference on Radioisotopes*, Tokyo, Japan Industrial Forum, Inc., 1969.

96. Harris, J. O., Eisenstark, A., and Dragsdorf, R. D., A study on the location of absorbed mercuric ions in *Escherichia coli*, *J. Bacteriol*, **68**:745, 1954.

97. Magos, L., Tuffery, A. A., and Clarkston, T. W., Volatilization of mercury by bacteria, *Brit. J. Industr. Med.*, **21**:294, 1964.

98. Westöö, G., Total mercury and methylmercury content in eggs bought in Sweden, June 1966–September 1967, *Var Foeda*, **19**:121, 1967; *Chem. Abstr.*, **73**:44027a, 1966.

99. Jensen, S., and Jernelöv, A., Biosyntes av metylkvicksilver. I. *Nordforsk Biocidinformation*, **10**:4, 1967.

100. Jensen, S., and Jernelöv, A., Biosyntes av metylkvicksilver. II Bildning av dimetylkvicksilver, *Nordforsk Biocidinformation*, **14**:3, 1968.

101. Jensen, S., and Jernelöv, A., Biological methylation of mercury in aquatic organisms, *Nature*, **223**:753, 1969.

102. Wood, J. M., A progress report on mercury, *Environment*, **14**:35, 1972.

103. Jernelöv, A., Release of methyl mercury from sediments with layers containing inorganic mercury at different depths, *Limnol. Oceanogr.*, **15**:958, 1970.

104. Anon., Mercury-laden sludge cannot be dumped, dredging at standstill, *Toronto Globe and Mail*, March 9, 1971.

105. Jernelöv, A., Mercury and food chains. In R. Hartung and B. D. Dinman, Eds., *Environmental Mercury Contamination*, Ann Arbor, Ann Arbor Science, 1972.

106. Fagerstrom, T., and Jernelöv, A., Some aspects of the quantitative ecology of mercury, *Water Res.*, **6**:1193, 1972.

107. Landner, L., Biochemical model for the biological methylation of mercury suggested from methylation studies *in vivo* with *Neurospora crassa*, *Nature*, **230**:452, 1971.

108. Blaylock, B. A., and Stadtman, T. C., Methane biosynthesis by *Methanosarcine barkeri*, *Arch. Biochem. Biophys.*, **116**:138, 1966.

109. Imura, N., Sukegawa, E., Pan, S-K., Nagao, K., Kim, J-Y., Kwan, T., and Ukita, T., Chemical methylation of inorganic mercury with methylcobalamin, a vitamin B_{12} analog, *Science*, **172**:1248, 1971.

110. Nelson, N., Byerly, T. C., Kolbye, A. C., Kurland, L. T., Shapiro, R. E.,

Shibko, S. I., Stickel, W. H., Thompson, J. E., Van Den Berg, L. D., and Weissler, A., Hazards of Mercury, *Environ. Res.*, **4**:1, 1971.

111. Lindstrand, L., Isolation of methylcobalamin from natural source material, *Nature*, **204**:188, 1964.

112. Brodie, J. D., Burke, G. T., and Mangum, J. H., Methylcobalamin as an intermediate in mammalian methionine biosynthesis, *Biochemistry*, **9**:4297, 1970.

113. Langley, D. G., Mercury methylation in an aquatic environment, *J. Wat. Pollut. Cont. Fed.*, **45**:44, 1973.

114. Wood, J. M., Environmental pollution by mercury. In J. N. Pitts and R. L. Metcalf, Eds., *Advances in Environmental Science and Technology*, Volume 2, New York, Wiley-Interscience, 1971.

115. Bishop, P. L., and Kirsch, E. J., Biological generation of methylmercury in anerobic pond sediment, *Proc. 27th Purdue Ind. Waste Conf., Lafayette*, **141**:628, 1972.

116. Furukawa, K., Suzuki, T., and Tonomura, K., Decomposition of organic mercurial compounds by mercury-resistant bacteria, *Agr. Biol. Chem. (Tokyo)*, **33**:128, 1969.

117. Dunlap, L., Mercury: Anatomy of a pollution problem, *Chem. Eng. News*, **49(27)**:22, 1971.

118. Bligh, E. G., Mercury and the contamination of freshwater fish, Winnipeg, Fisheries Research Board of Canada Manuscript Report No. 1088, 1970.

119. Wobeser, G., Nielsen, N. O., Dunlop, R. H., and Atton, F. M., Mercury concentrations in tissues of fish from the Saskatchewan River, *J. Fish. Res. Bd. Can.*, **27**:830, 1970.

120. Bligh, E. G., Mercury levels in Canadian fish. In *Mercury in Man's Environment*, Ottawa, Royal Society of Canada, 1971.

121. Anon., Preliminary report on a study of mercury as a pollutant in Ontario waters, Ontario Water Resources Commission Task Force Report, Toronto, Ontario Water Resources Commission, February 1970.

122. Branch, C. B., Testimony presented at the hearings before the Subcommittee on Energy, Natural Resources and the Environment of the Committee on Commerce, United States Senate, Ninety-First Congress, 2nd Session on *The Effects of Mercury on Man and the Environment*, Part 1, Serial 91-73, Washington, D.C., U.S. Government Printing Office, 1970.

123. Purdy, R., Testimony presented at the hearings before the Subcommittee on Energy, Natural Resources and the Environment of the Committee of Commerce, United States Senate, Ninety-First Congress, 2nd Session on *The Effects of Mercury on Man and the Environment*, Part 1, Serial 91-73, Washington, D.C., U.S. Government Printing Office, 1970.

124. Turney, W. G., Mercury pollution: Michigan's action program, *J. Wat. Pollut. Cont. Fed.*, **43**:1429, 1971.

125. Kolbye, A. C., Testimony presented at the Hearings before the Subcommittee on Energy, National Resources and the Environment of the Committee of Commerce, United States Senate, Ninety-First Congress, 2nd Session on *The Effects of Mercury on Man and the Environment*, Part 1, Serial 91-73, Washington, D.C., U.S. Government Printing Office, 1970.

126. Anon., Deadly mercury found in St. Clair fish, *The Detroit Free Press*, March 25, 1970.

127. Kutsuna, M., Historical perspective of the study on Minamata Disease. In *Minamata Disease*, Study Group of Minamata Disease, Kumamoto University, Japan, 1968.

128. The Mercury Problem—Symposium Concerning Mercury in the Environment, January 24–26, 1966, Stockholm. Proceedings Published as *Oikos*, **9 (Suppl)**, 1967.

129. Stickel, L. F., Discussion following Westöö, G., Methylmercury compounds in animal foods. In M. W. Miller and G. G. Berg, Eds., *Chemical Fallout*, Springfield, Thomas, 1969.

130. Johnels, A., Discussion following Westöö, G., Methylmercury compounds in animal foods. In M. W. Miller and G. G. Berg, Eds., *Chemical Fallout*, Springfield, Thomas, 1969.

131. Jernelöv, A., Factors in the transformation of mercury to methylmercury. In R. Hartung and B. D. Dinman, Eds., *Environmental Mercury Contamination*, Ann Arbor, Ann Arbor Science, 1972.

132. Berg, W., Johnels, A., Sjostrand, B., and Westermark, T., Mercury content in feathers of Swedish birds from the past 100 years, *Oikos*, **17**:71, 1966.

133. Hanson, A., Man-made source of mercury. In *Mercury in Man's Environment*, Ottawa, Royal Society of Canada, 1971.

134. Mitchell, R. L., Mercury consumption survey, April 10, 1969, The Chlorine Institute, 342 Madison Ave., New York, 1969.

135. Berlin, M., Discussion following Westöö, G., Methylmercury compounds in animal foods. In M. W. Miller and G. G. Berg, Eds., *Chemical Fallout*, Springfield, Thomas, 1969.

136. Noren, K., and Westöö, G., Methylmercury in fish, *Var Foeda*, **19**:13, 1967; *Chem. Abstr.*, **73**:44022v, 1970.

137. Jernelöv, A., Discussion following Westöö, G., Methylmercury compounds in animal foods. In M. W. Miller and G. G. Berg, Eds., *Chemical Fallout*, Springfield, Thomas, 1969.

138. Anon., The Mad Hatter Legacy, *Newsweek*, **75(16)**:72, 1970.

139. Anon., A fisherman says "poison has made him stronger." *Toronto Daily Star*, August 8, 1970.

140. Anon., We all feel healthy, *Kenora Daily Miner and News*, April 11, 1970.

141. McDuffie, B. R., How much mercury is too much? In *Proceedings of the Sixth Water Quality Symposium*, Washington, D.C., National Water Quality Research Council, April 18–19, 1972.

142. Evans, R. J., Bails, J. D., and D'Itri, F. M., Mercury levels in muscle tissues of preserved museum fish, *Environ. Sci. Tech.*, **6**:901, 1972.

143. Mayo, F., Testimony presented at the hearings before the Subcommittee on Energy, Natural Resources, and the Environment of the Committee on Commerce, United States Senate, Ninety-First Congress, 2nd Session on *The Effects of Mercury on Man and the Environment*, Part 1, Serial 91-73, Mt. Clemens, Michigan, May 8, 1970.

144. Anon., Report on the mercury pollution of the St. Clair River System, Toronto, Ontario Water Resources Commission, June 1970.

145. Fripp, B., Vermont fishing interest plunges following U.S. mercury contamination, *The Boston Evening Globe*, July 18, 1970.

146. Anon., Mercury pollution, *Sport Fish. Inst. Bull.*, **214**:6, 1970.

147. Anon., $$ values of fish, *Sport Fish. Inst. Bull.*, **227**:3, 1971.

148. Anon., Mercury pollution spreading, *Amer. Assoc. Professors Sanit. Eng. (AAPSE) Newsl.*, **5**:17, 1970.

149. Lambou, V. W., Problem of mercury emissions into the environment of the United States, Washington, D.C., U.S. Environmental Protection Agency, 1972.

150. Anon., Mercury pollution survey by FWQA, *Sport Fish. Inst. Bull.*, **221**:4, 1971.

151. Miller, G. E., Grant, P. M., Kishore, R., Steinkruger, F. J., Rowland, F. S., and Guinn, V. P., Mercury concentrations in museum specimens of tuna and swordfish, *Science*, **175**:1121, 1972.

152. Anon., Tuna may be checked across the U.S. by FDA for mercury levels, *The Wall Street Journal*, December 7, 1970.

153. Anon., FDA estimates 23% of canned tuna contaminated by excess mercury, *Air/Water Pollut. Rep.*, **8**:515, 1970.

154. Main, R. E., Mercury in the ecosystem, Unpublished paper, San Jose State College, San Jose, California, April 1971.

155. Anon., Mercury testing to cause higher priced canned tuna, *The Michigan State News*, May 24, 1971.

156. Anon., 843 Cases of tuna recalled, *The Detroit Free Press*, August 5, 1973.

157. Korms, R. F., The frustrations of Betty Russow, *Nutr. Today*, **7**:21, 1972.

158. Kahn, E., California Department of Public Health, Berkeley, Personal communication, 1974.

159. Anon., Mercury in swordfish, *Chem. Eng. News*, **49(20)**:11, 1971.

160. Smith, C., Mercury peril discounted at Scripps panel, *The San Diego Union*, June 3, 1971.

161. Hartman, W. L., Statement on Lake Erie. In *Proceedings of the Conference in the Matter of Pollution of Lake Erie and Its Tributaries*, Fifth Session, Detroit, Volume 1, 2, U.S. Department of Interior, Federal Water Quality Administration, 1970.

162. Lunde, G., Activation analysis of trace elements in fishmeal, *J. Sci. Food Agr.*, **19**:432, 1968.

163. Beasley, T. M., Mercury in selected fish protein concentrates, *Environ. Sci. Tech.*, **5**:634, 1971.

164. Gasiewicz, T. A., and Dinan, F. J., Concentration of mercury in the manufacture of fish protein concentrate by isopropyl alcohol extraction of Sheepshead and Carp., *Environ. Sci. Tech.*, **6**:726, 1972.

165. Lambou, V. W., Proposed Environmental Protection Agency Position Document—Mercury. Washington, D.C., U.S. Environmental Protection Agency, December 1971.

166. Newberne, P. M., Glaser, O., Friedman, L., and Stillings, B. R., Chronic exposure of rats to methylmercury in fish protein, *Nature*, **237**:40, 1972.

167. Anon., FDA allows mercury contaminated whale meal for pet use, *Food Chem. News*, **13**:30, 1971.

168. Anon., Fish kills. *Sport Fish. Inst. Bull.*, **221**:3, 1971.

169. Anon., Pollution fish kills in 1970. *Sport Fish. Inst. Bull.*, **241**:2, 1973.

170. Anon., *Fish Kills Caused by Pollution in 1972*. Environmental Protection Agency, Washington, D.C., U.S. Government Printing Office, 1972.

171. Anon., *Fish-kill in Boone Reservoir, July 9–13, 1968*, Chattanooga, Tennessee Valley Authority, Division of Health and Safety, Water Quality Branch, December 1968.

172. Toffler, A., *Future Shock*, New York, Bantam, 1971.

173. Anon., Dryden area residents object to publicity. *Kenora Daily Miner and News*, April 9, 1970.

174. Anon., Wabigoon water pollution beaten, claims Bernier, *The Toronto Telegram*, May 19, 1971.

175. Anon., Dryden firm stops mercury pollution but Ontario says danger remains, *The Toronto Daily Star*, July 23, 1970.

176. Anon., Dryden Chemicals established "61," *The Dryden Observer*, November 12, 1970.

177. Flewelling, F. J., Canadian experience with the reductions of mercury at chloralkali plants. In P. A. Krenkel, Ed., *Heavy Metals in the Aquatic Environment*, Oxford, Pergamon, 1975.

178. Watt, A. K., Deane, P., Dobson, B. R., Dodge, D. P., Neil, J. H., and

Stopps, G. J., Fourth report of the Mercury Task Force with summary of previous reports. Toronto, Toronto Ministry of the Environment, March 21, 1973.

179. Lee, B. L., Bottle of whiskey for a mercury warning sign. *The Hamilton Spectator*, May 3, 1972.

180. Anon., No title, Kenora District Camp Owners Association, *Vacation-land Camp Owners Trade J.*, **19(9)**:3, 1970.

181. Fimreite, N., and Reynolds, L. M., Mercury contamination of fish in northwestern Ontario, *J. Wildl. Manage.*, **37**:62, 1973.

182. Anon., Lake pollution 33 times safe level, *The Toronto Globe and Mail*, July 29, 1970.

183. Fimreite, N., Holsworth, W. N., Keith, J. A., Pearce, P. A., and Gruchy, I. M., Mercury in fish and fish-eating birds near sites of industrial contamination in Canada, *Can. Field-Nat.*, **85**:211, 1971.

184. Lee, B. L., No answers for lodge owners at mercury pollution seminar, *The Hamilton Spectator*, October 2, 1970.

185. Henry, J., It is no longer possible to be an Indian, *Maclean's (Canada)*, **84(6)**:47, 1971.

186. Takeuchi, T., School of Medicine, Kumamoto University, Japan, Personal communication, October 1972.

187. Burton, L., Indians press for inquest, *Kenora Daily Miner and News*, November 23, 1972.

188. Tsubaki, T., Clinical and epidemiological aspects of organic mercury intoxication. In *Mercury in Man's Environment*, Ottawa Royal Society of Canada, 1971.

189. Lee, B. L., Hair samples reveal 12 times normal mercury level, *The Hamilton Spectator*, June 2, 1970.

190. Lee, D., Indians threatened, *Winnipeg Free Press*, March 10, 1973.

191. Lamm, M., Gimli, Manitoba, Personal communication, September 1972.

192. Anon., Kenora birthrate highest but infant deaths soar, *Kenora Daily Miner and News*, September 23, 1971.

193. Anon., Minutes of a meeting held between the "Fish for Fun" Camp Owners Association and the Prime Minister of Ontario and the Honorable members of his cabinet, January 9, 1971.

194. Brunelle, R., and MacNee, W. Q., *Mercury in Fish*, Toronto, Ontario Department of Lands and Forests, May 1971.

195. Anon., Reed seeks compensation, *Kenora Daily Miner and News*, November 4, 1971.

196. Stopps, G. J., Medical Consultant, Health Studies Service, Department of Health, Toronto, Letter communication to Colin Myles, Doug Hook's Separation Lake Camp Ltd., Kenora, Ontario, February 28, 1972.

197. Sutherland, R. B., Chief of Health Studies Service, Department of

Health, Toronto, Letter communication to Colin Myles, Doug Hook's Separation Lake Camp Ltd., Kenora, Ontario, November 16, 1971.

198. Anon., The slow death of mercury poisoning, *Akwesasne Notes*, **7(3)**:16, 1975.

199. Derban, L. K. A., Outbreak of food poisoning due to alkylmercury fungicide on southern Ghana state farm, *Arch. Environ. Health*, **28**:49, 1974.

200. Ui, J., School of Urban Engineering, Tokyo, Japan, Personal communication, August 1975.

201. Gigon, F., The crime of Minamata, *L'Express (Paris)* No. 1201, July 15–21, 1974.

202. Kiyoura, R., Water pollution and Minamata Disease, *Int. J. Air Wat. Pollut.*, **7**:459, 1963.

203. Kiyoura, R., Water pollution and Minamata Disease, *Advan. Wat. Pollut. Res.*, **3**:291, 1964.

204. Moore, B., Discussion of R. Kiyoura's paper, *Advan. Wat. Pollut. Res.*, **3**:302, 1964.

205. Anon., The twisted face of Minamata, *Newsweek*, **81(14)**:38, 1973.

206. Ui, J., School of Urban Engineering, University of Tokyo, Tokyo, Japan, Personal communication, June 1976.

207. Anon., Frightened Japanese shun fish, *The Detroit Free Press*, July 8, 1973.

208. Kubota, Y., Struggle of Minamata victims '70–'75. Unpublished statement by the Minamata Indictment group, Tokyo, Translation by Jun Ui, University of Tokyo, Japan, 1975.

209. Anon., Commerce department considers economic aid for factory, and aiding one polluting industry would encourage requests by others for government aid, *Nihon Keizai*, January 23, 1975. Translation by Kikuji Saito.

210. Sneider, H., Environmental contamination by mercury compounds. *U.S. Department of State Airgram, Tokyo AO751*, July 24, 1970.

211. D'Itri, F. M., and D'Itri, P. A., The distribution of mercury in fish and its form of occurrence—a discussion. In P. A. Krenkel, Ed., *Heavy Metals in the Aquatic Environment*, Oxford, Pergamon, 1975.

212. Smith, W. E., and Smith, A., Death from the water. In *Minamata, Words and Photographs*, New York, Holt, Rinehart and Winston, 1975.

213. Harada, M., Fujino, T., Akagi, T., and Nishigaki, S., Epidemiological and clinical study and historical background of mercury pollution on Indian Reservations in northwestern Ontario, Canada, *Bull. Inst. Const. Med. (Kumamoto Univ.)*, **26(3/4)**:169, 1976.

214. Takeuchi, T., D'Itri, F. M., Fischer, P. V., Annett, C. S. and Okabe,

M., The outbreak of Minamata Disease (methyl mercury poisoning) in cats on northwestern Ontario Reserves, *Environ. Res.,* In press.

215. Ui, J., Intermediate report on Indians in Ontario, August 17, 1975, University of Tokyo, Toyko, Japan, Unpublished report.

216. West, J. M., Mercury. In A. E. Schreck, Ed., *Minerals Yearbook 1971,* U.S. Department of the Interior, Bureau of Mines, Washington, D.C., U.S. Government Printing Office, 1971.

217. Oliver, T., Examination of workers in dangerous trades, *Brit. Med. J.,* 2:745, 1902.

218. Lee, W. R., The history of the statutory control of mercury poisoning in Great Britain. *Brit. J. Industr. Med.,* 25:52, 1968.

219. Smith, A. D. M., and Miller, J. W., Treatment of inorganic mercury poisoning with N-acetyl-D, L-penicillamine, *Lancet,* 1:640, 1961.

220. Pennington, J. W., Mercury: A materials survey. Bureau of Mines Information Circular No. 7941, Washington, D.C., U.S. Government Printing Office, 1959.

221. Engel, G. T., Mercury. In H. F. Maek, J. J. McKetta, Jr., and D. F. Othmer, Eds., *Encyclopedia of Chemical Technology,* second ed., Vol. 13, New York, Interscience, 1966.

222. Anon., List of mercury polluters and status of government action to halt mercury discharge. Environmental Protection Agency, Appendix 2. In Hearing before a Subcommittee of the Committee on Government Operations, House of Representatives, Ninety-Second Congress, 1st. Session. *Mercury Pollution and Enforcement of the Refuse Act of 1899,* Part 1, Washington, D.C., U.S. Government Printing Office, 1971.

223. Berlin, M., Clarkson, T. W., Friberg, L. T., Gage, J. C., Goldwater, L. J., Jernelöv, A., Kazantzis, G., Magos, L., Nordberg, G. F., Rodford, E. P., Ramel, C., Skerfving, S., Smith, R. G., Suzuki, T., Swensson, A., Tejning, S., Fruhaut, R., and Vostal, J., Maximum allowable concentrations of mercury compounds, *Arch. Environ. Health,* 19:891, 1969.

224. Johnson, H. R. M., and Koumides, O., Unusual case of mercury poisoning, *Brit. Med. J.,* 1:340, 1967.

225. Lambert, S. W., and Patterson, H. W., Poisoning by mercuric chloride and its treatment, *Arch. Intern. Med.,* 16:865, 1915.

226. Berger, S. S., Applebaum, H. S., and Young, A. M., Immediate cecostomy and constant lavage in mercuric chloride poisoning, *J. Amer. Med. Assoc.,* 98:700, 1932.

227. Skerfving, S., and Vostal, J., Symptoms and signs of intoxication. In L. Friberg and J. Vostal, Eds., *Mercury In the Environment,* Cleveland, CRC, 1972.

228. Stahl, Q. R., Preliminary air pollution surveys of mercury and its compounds. U.S. Department of Health, Education and Welfare, Public

Health Service, National Air Pollution Control Administration, Raleigh, N.C., 1969.

229. Anon., Threshold limit values of airborne contaminants for 1968: recommended and intended values. Adopted at the American Conference of Governmental Industrial Hygienists. St. Louis, Mo., 1968.

230. Anon., Threshold limit values of airborne contaminants and physical agents with intended changes. Americna Conference of Governmental Industrial Hygienists. Cincinnati, Ohio, 1971.

231. Anon., National emission standards for hazardous air pollutants—proposed standards for asbestos, beryllium, mercury, *Fed. Regist.*, **36**: 23239, 1971.

232. Anon., Occupational safety and health standards, *Fed. Regist.*, **36**: 15101, 1971.

233. Neal, P. A., Mercury poisoning from the public health viewpoint, *Amer. J. Public Health*, **28**:907, 1938.

234. Jonasson, I. R., and Boyle, R. W., Geochemistry of mercury and origins of natural contaminations of the environment, *Can. Inst. Mining Met. Bull.*, **65**:32, 1972.

235. Bateman, A. M., *Economic Mineral Deposits*, second ed., New York, Wiley, 1955.

236. Goldwater, L. J., *Mercury: A History of Quicksilver*, Baltimore, York, 1972.

237. Hamilton, A., and Hardy, H., *Industrial Toxicology*, New York, Harper, 1949.

238. Vauk, V. B., Fugas, M., and Topolnik, Z., Environmental conditions in the mercury mine of Idria, *Brit. J. Industr. Med.*, **7**:168, 1950.

239. Henderson, Y., and Haggard, H. W., Inorganic and organometallic gases. In *Noxious Gases and the Principles of Respiration Influencing their Action*, ACS Monograph, No. 35, second ed., New York, Reinhold, 1943.

240. Byrne, A. R., and Kosta, L., Studies on the distribution and uptake of mercury in the area of the mercury mine at Idrija, Slovenia (Yugoslavia), *Vestn. Slov. Kem. Drus.*, **17**:5, 1970; *Chem. Abstr.*, **74**:130020v, 1971.

241. Weeks, M. E., and Leicester, H. M., *Discovery of the Elements*, seventh Ed., Easton, Pa., American Chemical Society, 1968.

242. Barba, A. A., *El Arte de los Metales*. English translation by R. E. Douglass and E. P. Mathewson, New York, Wiley, 1923.

243. Flawn, P. T., Vice President for Academic Affairs, University of Texas-Austin, Personal communication, 1970.

244. Downer, S. A., On her trip into the New Almaden mine, 1854, *Calif. J. Mines Geol.*, **49**:113, 1953.

245. Forbes, J. A., Examination of James Alexander Forbes, resumed December 15th, 1857, *Calif. J. Mines Geol.*, **49**:87, 1953.

246. Lyman, C. S., On mines of cinnabar in upper California, 1848, *Calif. J. Mines Geol.*, **49**:109, 1953.

247. Holmes, J., Mercury is heavier than you think, *Esquire*, **75(5)**:135, 1971.

248. Anon., Mercury in the California environment. Interim report of the Interagency Committee on Environmental Mercury, Berkeley, California State Department of Health, July 1971.

249. Lord, J. K., On his trip into the New Almaden mine, 1860, *Calif. J. Mines Geol.*, **49**:125, 1953.

250. Anon., *Mineral Resources of the United States, 1902*. U.S. Geological Service, Washington, D.C., U.S. Government Printing Office, 1904.

251. Rentos, P. G., and Seligman, E. J., Relationship between environmental exposure to mercury and clinical observation, *Arch. Environ. Health*, **16**:794, 1968.

252. Buckell, M., Hunter, D., Milton, R., and Perry, K. M. A., Chronic mercury poisoning, *Brit. J. Industr. Med.*, **3**:55, 1946.

253. Grant, N., Mercury in man. *Environment*, **13**:3, 1971.

254. Stahl, Q. R., *Preliminary Air Pollution Survey of Mercury and Its Compounds*. U.S. Department of Health, Education and Welfare, Public Health Service, National Air Pollution Control Administration, Raleigh, N.C., 1969.

255. Dodgson, C. L. (L. Carroll), *Alice's Adventures in Wonderland*, New York, Random House, 1946.

256. Smith, G., *The Concise Dictionary of National Biography, Part I*, London, Oxford, 1961.

257. Porter, C., Felt hat making: Its processes and hygiene, *Brit. Med. J.*, **1**:377, 1902.

258. Taylor, J. G., Chronic mercurial poisoning with special reference to the dangers in hatters' furriers' manufactories. In J. H. Bryan and F. J. Stewart, Eds., *Guy's Hospital Reports* (Volume XL, third series), London, J. & A. Churchill, 1901.

259. Neal, P. A., Flinn, R. H., Edwards, T. I., Reinhart, W. H., Hough, J. W., Dallovale, J. M., Goldman, F. H., Armstrong, D. W., Gray, A. S., Coleman, A. L., and Postman, B. F., Mercurialism and its control in the felt-hat industry. *U.S. Public Health Bulletin No. 263*, 1941.

260. Hamilton, A., Industrial diseases of fur cutters and hatters, *J. Ind. Hyg.*, **4**:219, 1922.

261. Lee, W. R., The history of the statutory control of mercury poisoning in Great Britain, *Brit. J. Industr. Med.*, **25**:52, 1968.

262. Neal, P. A., Jones, R. R., Bloomfield, J. J., Dallavalle, J. M., and Edwards, T. I., A study of chronic mercurialism in the hatters' fur-cutting industry. *U.S. Public Health Bull. No. 234*, 1937.

263. Tylecote, F. E., Remarks on industrial mercurial poisoning as seen in felt-hat makers, *Lancet*, **2**:1137, 1912.

264. Faulds, H., On the skin-furrows of the hand, *Nature*, **22**:605, 1880.

265. Herschel, W. J., Skin furrows of the hand, *Nature*, **23**:76, 1880.

266. Faulds, H., On the identification of habitual criminals by fingerprints, *Nature*, **50**:548, 1894.

267. Agate, J. N., and Bucknell, M., Mercury poisoning from fingerprint photography: an occupational hazard of policemen, *Lancet*, **2**:45, 1949.

268. Anon., Fingerprinters poisoned. *Sci. News Lett.*, **56**:231, 1949.

269. Budge, E. A. W., *The Nile*, Thomas Cook and Son (Egypt), Ltd., London, 1912.

270. Anon., Trends in the usage of mercury. National Materials Advisory Board, Report No. AD693873, Washington, D.C., National Research Council, 1969.

271. Erich, E., Precision is key in making bowling balls, *Rubber World*, **162(4)**:67, 1970.

272. Aaronson, T., Mercury in the environment. *Environment*, **13**:16, 1971.

273. Noe, F. E., Mercury as a potential hazard in medical laboratories. *New Eng. J. Med.*, **261**:1002, 1959.

274. Fehr, F., *The Use and Health Hazard of Mercury in Saskatchewan* Part II—Laboratories, Hospitals, and Dental Offices, Occupational Health and Safety Division, Saskatchewan Department of Labour, December 1973.

275. Lomonosov, M. V., *Complete Collected Works of Lomonosov*, Moscow and Leningrad, Russian Academy of Sciences, 1952.

276. Copplestone, J. F., and McArthur, D. A., An inorganic mercury hazard in the manufacture of artificial jewellery, *Brit. J. Ind. Med.*, **24**:77, 1967.

277. Anon., Mercury ingested from broken thermometers or as a water pollutant—what hazard? *J. Amer. Med. Assoc.*, **215**:647, 1971.

278. Ryrie, D. R., Toghill, P. J., Tanna, M. K., and Galan, G. H., Marrow suppression from mercury poisoning? *Brit. Med. J.*, **1**:499, 1970.

279. Gorton, B., Mercurial poisoning, *Vet. J.*, **80**:49, 1924.

280. Cooke, N. E., and Beitel, A., Some aspects of other sources of mercury to the environment. In *Mercury in Man's Environment*, Ottawa, Royal Society of Canada, 1971.

281. Anon., Mercury exposure found in school laboratories, *Ind. Hyg. Newsl.*, **8**:8; 1971.

282. Jordan, L., and Barrows, W. P., Mercury poisoning from electric furnaces, *Ind. Eng. Chem.*, **16**:898, 1924.

283. Goldwater, L. J., Kleinfeld, M., and Berger, A. R., Mercury exposure in a university laboratory, *Arch. Industr. Health*, **15**:245, 1956.

284. Anon., Mercury, not mum? *New Sci. Sci. J.*, **52**:274, 1971.

285. Stock, A., and Cucuel, F., Die verbreitung des quecksilbers, *Naturwiss,* **22**:390, 1934.

286. Dennis, L. M., Poisoning by mercury vapor, *Ind. Eng. Chem.,* **18**:1205, 1926.

287. Frey, J. E., Alfred Stock—mercurial chemist. Paper presented at 75th Annual Meeting of Michigan Academy of Science, Arts and Letters, Western Michigan University, Kalamazoo, Michigan, April 1971.

288. Stock, A., Danger from mercury vapor, *Z. angew. Chem.,* **39**:461, 1926; *Chem. Abstr.,* **20**:2214, 1926 (in German).

289. Giese, A. C., Mercury poisoning, *Science,* **91**:476, 1940.

290. Stock, A., The handling of mercury, *Z. angew. Chem.,* **42**:999, 1926; *Chem. Abstr.,* **24**:669, 1930 (in German).

291. Winderlich, V. R., Eine Vergiftung durch Quecksilberdampf vor 150 Jahren, *Chemiker-Ztg.,* **52**:29, 1928.

292. Anon., Mercury, properties and uses. In W. Haley, Ed., *Encyclopaedia Britannica,* London, William Benton, 1969.

293. Anon., *Mercury Quarterly.* Mineral Industries Surveys, Washington, D.C., U.S. Department of the Interior, 1950–1975.

294. Anon., *Manufacture of Industrial Chemicals.* Statistics Canada, Manufacturing and Primary Industries Division, Ottawa, Information Canada, 1973.

295. Walker, H. J., Industrial mercurial poisoning: With notes of two cases, *Lancet,* **2**:823, 1905.

296. Bidstrup, P. L., Bonnell, J. A., Harvey, D. G., and Locket, S., Chronic mercury poisoning in man repairing direct-current meters, *Lancet,* **2**:856, 1951.

297. Lewis, L., Mercury poisoning in tungsten molybdenum rod and wire manufacturing industry, *J. Amer. Med. Assoc.,* **129**:123, 1945.

298. Kazantzis, G., Chronic mercury poisoning, clinical aspects, *Ann. Occup. Hyg.,* **8**:65, 1965.

299. Curley, E., Mercury poisoning linked to ITT laxity, *Jt. Issue,* **3**:3, 1972.

300. Williams, C. R., Eisenbid, M., and Pihl, S. E., Mercury exposures in dry battery manufacture, *J. Industr. Hyg. Toxicol.,* **29**:378, 1947.

301. Tamis, M., Bornstein, B., Bebar, M., and Chwat, M., Mercury poisoning from an unsuspected source, *Brit. J. Industr. Med.,* **21**:299, 1964.

302. Klein, D. H., Statement at hearings before the Subcommittee on Energy, Natural Resources, and the Environment of the Committee on Commerce. United States Senate, Ninety-First Congress, 2nd Session, on *The Effect of Mercury on Man and the Environment,* Part 2, Serial No. 91–73, Washington, D.C., U.S. Printing Office, 1970.

303. Benning, D., Outbreak of mercury poisoning in Ohio, *Industr. Med. Sur.,* **27**:354, 1958.

304. Flewelling, F. J., Loss of mercury to the environment from chloralkali plants. In *Mercury in Man's Environment*, Ottawa, Royal Society of Canada, 1971.

305. Anon., Harsh words for Mercury, *Chemistry in Canada,* June 1970.

306. Anon., Chlorine. In W. Haley, Ed., *Encyclopaedia Britannica*, London, William Benton, 1969.

307. Anon., *North American Chloralkali Industry Plants and Production Data Book*. Pamphlet No. 10, The Chlorine Institute, 342 Madison Ave., New York, 1973.

308. Anon., Chlorine gas, *Chemicals*, **17(2)**:35, 1970.

309. Friberg, L., Hammarstrom, S., and Hystrom, A., Kidney injury after chronic exposure to inorganic mercury, *Arch. Ind. Hyg. Occup. Med.*, **8**:149, 1953.

310. Hernberg, S., and Hasanen, E., Relationship of inorganic mercury in blood and urine, *Work—Environmental—Health*, **8**:39, 1971.

311. Smith, R. G., Vorwald, A. J., Patil, L. S., and Mooney, T. F., Effects of exposure to mercury in the manufacture of chlorine, *Amer. Ind. Hyg. Assoc. J.*, **31**:687, 1970.

312. Anon., Chlorine production drops as demand lags, *Chem. Eng. News*, **49(46)**:10, 1971.

313. Wallace, R. A., Fulkerson, W., Shults, W. D., and Lyon, W. S., *Mercury in the Environment, The Human Element*, Oak Ridge, Oak Ridge National Laboratory, 1971.

314. Sommers, H. A., The chloro-alkali industry, *Chem. Eng. Progr.*, **61**:94, 1965.

315. Fritsch, A. J., Testimony during hearings before the Subcommittee on Energy, Natural Resources, and the Environment of the Committee on Commerce. United States Senate, Ninety-first Congress, 2nd Session, On *The Effect of Mercury on Man and the Environment*, Part 2, Serial No. 91-73, Washington, D.C., U.S. Government Printing Office, 1970.

316. Anon., Process controls mercury emissions, *Chem. Eng. News*, **50(8)**:17, 1972.

317. Anon., Environmental Protection Agency (PR Notice 72-5) Certain Products Containing Mercury, Cancellation of Registration, March 22, 1972. In *Fed. Regist.*, **37(61)**:6119, 1972.

318. Goldwater, L. J., and Jeffers, C. P., Mercury poisoning from the use of anti-fouling plastic paint, *J. Ind. Hyg. Toxicol.*, **24**:21, 1942.

319. Novack, S., Mercury in the environment, *Environment*, **11**:3, 1969.

320. Hoffman, E., Determination of phenylmercuric compounds and total mercury in paints, *Z. Analyt. Chem.*, **182**:193, 1961.

321. Anon., Crabbiness, insomnia—due to mercury in latex paint? *Res. Develop.*, **23**:8, 1972.

322. Griffith, W. H., Mercury contamination in California's fish and wildlife. In D. R. Buhler, Ed., *Mercury in the Western Environment*, Corvallis, Continuing Education Publications, 1973.

323. Buhler, D. R., Claeys, R. R., and Rayner, H. J., Seasonal variations in mercury contents of Oregon pheasants. In D. R. Buhler, Ed., *Mercury in the Western Environment*, Corvallis, Continuing Education Publications, 1973.

324. Hirschman, S. Z., Feingold, M., and Boylen, G. W., Mercury in house paint as a cause of acrodynia, *New Eng. J. Med.*, **269**:889, 1963.

325. Hatfield, I., Further experiments with chemicals suggested as possible wood preservatives, *Proc. Amer. Wood Preservers' Assoc.*, 330, 1932; *Chem. Abstr.*, **27**:2553, 1933.

326. Vintinner, F. J., Dermatitis venenata resulting from contact with an aqueous solution of ethylmercury phosphate, *Ind. Hyg. Toxicol. J.*, **22**:297, 1940.

327. McCord, C. P., Meek, S. F., and Neal, T. A., Phenylmercuric oleate, skin irritant properties, *Ind. Hyg. Toxicol. J.*, **23**:466, 1941.

328. Ahlmark, A., Poisoning by methylmercury compounds, *Brit. J. Ind. Med.*, **5**:117, 1948.

329. Bouveng, H. O., The chlorine industry and the mercury problem, *Modern Kemi*, **3**:45, 1968 (in Swedish).

330. Paavila, H. W., Use of mercury in the Canadian pulp and paper industry. In *Mercury in Man's Environment*, Ottawa, Royal Society of Canada, 1971.

331. Anon., Quebec mercury poisoning sends 4 Indians to hospital, *Toronto Daily Star*, June 16, 1971.

332. Ruckelshaus, W. D., Certain products containing mercury, cancellation of registrations, *Fed. Regist.*, **37**:6419, 1972.

333. States, E. B., Comptroller General, Environmental Protection Agency. Efforts to remove hazardous pesticides from the channels of trade. Report to Congress, April 26, 1973.

334. Anon., Mercury reconsiderations asked by manufacturers, *Food Chem. News*, May 1, 1972.

335. Anon., Prehearing conference set on mercury cancellations, *Pesticide Chem. News*, **1**:3, 1973.

336. Anon., Manufacturers of mercury compounds object to cancellations, *Pesticide Chem. News*, **1**:8, 1973.

337. Anon., Cancellation of all mercury pesticides uses sought by EPA, *Pesticide Chem. News*, **1**:6, 1973.

338. Kevorkian, J., Cento, D. P., Hyland, J. R., Bagozzi, W. M., and van Hollebeke, E., Mercury content of human tissues during the Twentieth Century, *Amer. J. Public Health*, **62**:504, 1972.

339. Anon., Environmental Quality. First Annual Report of the Council on Environmental Quality. Washington, D.C., U.S. Government Printing Office, 1970.

340. Anon., Spectrum, *Environment,* **12**:51, 1970.

341. Klein, D. H., Department of Chemistry, Hope College, Holland, Michigan, Personal communication, 1973.

342. Joensuu, O. L., Fossil fuels as a source of mercury pollution, *Science,* **172**:1027, 1971.

343. Kennedy, E. J., Ruch, R. R., Gluskoter, N. J., and Shimp, N. F., Environmental studies of mercury and other sediments in coal and lake sediments as determined by neutron activation analysis. In J. R. Vogt, T. F. Parkinson, and R. L. Carter, Eds., *Nuclear Methods in Environmental Research,* University of Missouri, Columbia, 1971.

344. White, D. E., Mercury and base-metal deposits with associated thermal and mineral waters, *Geochemistry of Hydrothermal Ore Deposits,* New York, Holt, Rinehart, and Winston, 1967.

345. Shacklette, H. T., Boerngen, J. G., and Turner, R. L., Mercury in the environment surficial materials of the conterminous United States. Geological Survey Circular 644, U. S. Geological Survey, Washington, D.C., 1971.

346. Klein, D. H., and Goldberg, E. D., Mercury in the marine environment, *Environ. Sci. Technol.,* **4**:765, 1970.

347. Oertine, K. K., and Goldberg, E. D., Fossil fuel combination and the major sedimentary cycle, *Science,* **173**:233, 1971.

348. Anon., USDA soils laboratory studies mercury uptake and release, *Aminoco Lab News,* **29**:9, 1973.

349. Rouelle, H-M., Observations chymiques, *J. Med. (Chirurgie, Pharmacie, Ec.)*, **48**:322, 1777.

350. Proust, J-L., On the existence of mercury in the water of the ocean, *J. Phys.,* **49**:153, 1799; *Mem. Mus. Hist. Nat.,* **7**:479, 1821.

351. Garrigou, F., Sur la presence du mercure dans la source du Rocher, a l'establissement du mont Cornadore (Saint-Nectaire-le-Haut, Puy-de-Dome), *Comp. rend.,* **84**:963, 1887.

352. Willm, E., Sur la presence du mercure dans les eaux minerales de Saint-Nectaire, *Comp. rend.,* **88**:1032, 1879.

353. Bardet, J., Etude spectrographique des eaux minerales francaises, *Comp. rend.,* **157**:224, 1913.

354. Hammerstrom, R. J., Hissong, D. E., Koppler, F. C., Meyer, J., McFarren, E. F., and Pringle, B. H., Mercury in drinking water supplies, *J. Amer. Water Works Assoc.,* **64**:61, 1972.

355. Burton, J. D. and Leatherman, T. M., Mercury in a coastal marine environment, *Nature,* **231**:440, 1971.

356. Anon., First estuarian mercury study, *The Nation's Health (APHA),* August 1971.

357. Takeuchi, T., Department of Pathology, Kumamoto University, Kumamoto, Japan, Personal communication, 1970.

358. Smith, J. D., Nicholson, R. A., and Moore, P. J., Mercury in sediments from the Thames estuary, *Environ. Pollut.,* 4:153, 1973.

359. Almkvist, J., Some notes on the history of mercury intoxication, *Acta Med. Scand.,* 70:464, 1929.

360. Dioscorides, P., On the nature of cinnabar and quicksilver, First Century A.D. Translation by John Goodyear, 1655. In E. L. Egenhoff, *Calif. J. Mines Geol.,* 49:21, 1953.

361. Rosen, G., *The History of Miner's Disease,* New York, Schuman, 1943.

362. Katz, D., Why 430 blacks with syphilis went uncured for 40 years, *Detroit Free Press,* November 5, 1972.

363. Cleugh, J., *Secret Enemy: The Story of a Disease,* New York, T. Yoseloff, 1956.

364. Dennie, C. C., *A History of Syphilis,* Springfield, Thomas, 1962.

365. Holcomb, R. C., The Holy Wood and the Haitian myth of the origin of syphilis. In C. Butler, Ed., *Syphilis,* Lancaster, Science Press, 1938.

366. Baulch, P. W. F., *Mercury Pollution.* Ph.D. dissertation, Department of Mechanical Engineering, University of Toronto, 1972.

367. Cember, H., A model for the kinetics of mercury elimination, *Amer. Industr. Hyg. Assoc. J.,* 30:367, 1969.

368. Anon., JAMA 75 years ago. Mercury treatment for syphilis, *J. Amer. Med. Assoc.,* 215:549, 1971.

369. Afonso, J. F., and de Alvarez, R. R., Effects of mercury on human gestation, *Amer. J. Obstet. Gynec.,* 80:145, 1960.

370. Stokes, J. H., *The Third Great Plague: A Discussion of Syphilis for Everyday People,* Philadelphia, W. B. Saunders, 1920.

371. McGeorge, J. R., Mercurial stomatitis, *J. Amer. Dent. Assoc.,* 22:60, 1935.

372. Anon., Veneral diseases. In W. Yust, Ed., *Encyclopedia Britannica,* New York, William Benton, 1956.

373. Millar, A. F. W., Perchloride of mercury poisoning by absorption from the vagina, *Brit. Med. J.,* 2:453, 1916.

374. Anon., A new drug for both syphilis and leprosy, *Ind. Eng. Chem.,* 19:634, 1927.

375. Biskind, L. H., Phenylmercury nitrate. Its clinical uses in gynecology, a preliminary report, *Surg. Gynec. Obstet.,* 57:261, 1933.

376. Keith, N. M., and Whelan, M., A study of the action of ammonium chlorid and organic mercury compounds, *J. Clin. Invest.,* 3:149, 1926.

377. Batterman, R. C., DeGraff, A. C., Rose, O. A., Treatment of congestive heart failure with an orally administered mercurial diuretic, *Amer. Heart J.*, **21**:98, 1941.

378. Vogl, A., The discovery of the organic mercurial diuretics, *Amer. Heart J.*, **39**:881, 1950.

379. Handley, C. H., Diuretics. In V. A. Drill, Ed., *Pharmacology in Medicine*, New York, McGraw-Hill, 1954.

380. Crawford, J. H., and McIntosh, J. F., Observations on the use of Novosurol in edema due to heart failure, *J. Clin. Invest,* **1**:333, 1924.

381. DeGraff, A. C., and Hadler, J. E., A review of the toxic manifestations of mercurial diuretics in man, *J. Amer. Med. Assoc.*, **119**:1006, 1942.

382. Evans, H., and Perry, K. M. A., Immediate death after the use of intravenous mercurial diuretics, *Lancet*, **1**:576, 1943.

383. Vogl, A., The use of mercurial diuretics in congestive heart failure, *Mod. Concepts Cardiovasc. Dis.*, **24**:263, 1955.

384. Brown, E. H., Reactions to the organomercurial compounds, *Ann. NY Acad. Sci.*, **65**:550, 1957.

385. Stewart, H. J., McCow, H. I., Shepard, E. M., and Luckey, E. H., Experience with Thiomerin, a new mercurial diuretic, *Circulation*, **1**:502, 1950.

386. Best, M. M., Hurt, W. F., Shaw, J. E., and Wathen, J. D., Study of the mercurial diuretic, dicurin procaine (merethoxylline procaine) by subcutaneous injection, *Amer. J. Med. Sci.*, **225**:132, 1953.

387. Silverman, J. J., and Worthen, J. F., Agranulosis in a patient treated with mercurial diuretics, *J. Amer. Med. Assoc.,* **148**:200, 1952.

388. Hyman, H. T., Sudden death after use of mercurial diuretics, *J. Amer. Med. Assoc.*, **119**:1001, 1942.

389. Poll, D., and Stern, J. E., Dangers of dehydration treatment in heart diseases, *Med. Clin. North Amer.*, **21**:1873, 1937.

390. Pitts, R. F., Some reflections on mechanisms of action of diuretics, *Amer. J. Med.*, **24**:745, 1958.

391. Jones, J. W., School of Human Medicine, Michigan State University, East Lansing, Michigan, Personal communication, 1974.

392. Clarkson, T. W., Biochemical aspects of mercury poisoning, *J. Occup. Med.*, **10**:351, 1968.

393. Weston, R. E., The mode and mechanism of mercurial diuresis in normal subjects and edematous cardiac patients, *Ann. NY Acad. Sci.*, **65**:576, 1957.

394. Jameson, R. L., The action of a mercurial diuretic on active sodium transport electrical potential and permeability to chloride of the isolated toad bladder, *J. Pharmacol. Exp. Therap.*, **133**:1, 1966.

395. Blau, M., and Bender, M. H., Radiomercury (Hg-203) labeled Neohydrin. A new agent for brain tumor localization, *J. Nucl. Med.*, **3**:83, 1962.

396. Warkany, J., and Hubbard, D. M., Adverse mercurial reactions in the form of acrodynia and related conditions, *Amer. J. Dis. Child.*, **81**:335, 1951.

397. Petren, K., L'acrodynie: Une intoxication arsenicale, *Rev. Neurol.*, **37**: 812, 1921.

398. Pehu, M., and Ardisson, P., Une maladie qui ressuscite: L'acrodynie, *Paris Med.*, **1**:341, 1927.

399. Popkin, R. J., Calomel, *J. Amer. Med. Assoc.*, **216**:1347, 1971.

400. Underwood, A. L., Mercurial gangrene, *Cincinnati Lancet and Observer*, **8**:585, 1865.

401. Warkany, J., and Hubbard, D. M., Mercury in the urine of children with acrodynia, *Lancet*, **1**:829, 1948.

402. Wilson, V. K., Thomson, M. L., and Holzel, A., Mercury nephrosis in young children with special reference to teething powders containing mercury, *Brit. Med. J.*, **1**:358, 1952.

403. Farquhar, H. G., Mercurial poisoning in early childhood, *Lancet*, **2**:1186, 1953.

404. Bivingad, L., and Lewis, G., Acrodynia: A new treatment with BAL, *J. Pediat.*, **32**:65, 1948.

405. Pennington, J. W., Mercury: A materials survey. *Bureau of Mines Information Circular No. 7941*. U.S. Government Printing Office, Washington, D.C., 1959.

406. Hosford, G. N., and McKenney, J. P., Ointment of yellow mercuric oxide (Pagenstrecher's Ointment), *J. Amer. Med. Assoc.*, **100**:17, 1933.

407. Abramowicz, I., Deposition of mercury in the eye, *Brit. J. Ophthal.*, **30**:696, 1946.

408. Wheeler, M., Discoloration of the eyelids from prolonged use of ointments containing mercury, *Amer. J. Ophthal.*, **31**:441, 1948.

409. Goodman, L. S., and Gilman, A., *The Pharmacological Basis of Therapeutics*, New York, Macmillan, 1965.

410. Underwood, G. B., Gaul, I. E., Collins, E., and Mosby, M., Overtreatment dermatitis of the feet, *J. Amer. Med. Assoc.*, **130**:249, 1946.

411. Deakin, S., Unusual symptoms following application of unguentum hydrargyri ammoniati, *Brit. Med. J.*, **11**:1281, 1883.

412. Harper, P., Idiosyncrasy to ammoniated mercury ointment, *J. Pediat.*, **5**:794, 1934.

413. Stoneman, M. E. R., Pink disease after application of mercury ointment, *Lancet*, **1**:938, 1958.

414. Tutt, J. F. D., Mercurial poisoning in a dog, *Vet. J.*, **81**:559, 1925.

415. Stevens, G. G., Mercurial poisoning, *Cornell Vet.*, **28**:50, 1938.

416. Jones, H. R., *Mercury Pollution Control*, Park Ridge, New Jersey, Noyes Data Corporation, 1971.
417. Goeckermann, W. H., A peculiar discoloration of the skin, *J. Amer. Med. Assoc.*, **79**:605, 1922.
418. Goeckermann, W. H., A peculiar discoloration of the skin, *J. Amer. Med. Assoc.*, **84**:506, 1925.
419. Hollander, L., and Baer, H. L., Discoloration of the skin due to mercury, *Arch. Derm. Syph.*, **20**:27, 1929.
420. Lamar, L. M., and Bliss, B. O., Localized pigmentation of the skin due to topical mercury, *Arch. Derm.*, **93**:450, 1966.
421. Barr, R. D., Rees, P. H., Cordy, P. E., Kungu, A., Woodger, B. A., and Cameron, H. J., Nephrotic syndrome in adult Africans in Nairobi, *Brit. Med. J.*, **2**:131, 1972.
422. Novey, F. G., A generalized mercurial (cinnabar) reaction following tattooing, *Arch. Derm. Syph.*, **49**:172, 1944.
423. Ravits, H. G., Allergic tattoo granuloma, *Arch. Derm.*, **86**:287, 1962.
424. Biro, L., Unusual complications of mercurial (cinnabar) tattoo, *Arch. Derm.*, **96**:165, 1967.
425. McKenna, R. M. B., Allergy to tattooing, *Practitioner*, **160**:471, 1948.
426. Eagle, H., The effect of sulfhydryl compounds on the antispirochetal action of arsenic, bismuth, and mercury compounds in vitro, *J. Pharm. Exp. Ther.*, **66**:436, 1939.
427. Fildes, P., The mechanism of the anti-bacterial action of mercury, *Brit. J. Exp. Pathol.*, **21**:67, 1940.
428. Brewer, J. H., The antibacterial effects of organic mercurial compounds, *J. Amer. Med. Assoc.*, **112**:2009, 1939.
429. Bucker, W. H., Two unusual cases of mercury poisoning, *U.S. National Clearing House Poison Control Bulletin*, March/April 4, 1963.
430. Grant, N., Legacy of the Mad Hatter, *Environment,* **11**:18, 1969.
431. Russell, C. L., D.D.S., St. Charles, Michigan, Personal communication, 1972.
432. Garfield, S., *Teeth, Teeth, Teeth*, New York, Simon and Schuster, 1969.
433. Hoover, A. W., and Goldwater, L. J., Absorption and excretion of mercury in man X. Dental amalgams as a source of urinary mercury, *Arch. Environ. Health*, **12**:506, 1966.
434. Nixon, G. S. and Smith, H., Hazard of mercury poisoning in the dental surgery, *J. Oral Ther. Pharm.*, **1**:512, 1965.
435. Buchwald, L. H., Exposure of dental workers to mercury, *Amer. Ind. Hyg. Assoc. J.*, **33**:429, 1972.
436. Gronka, P. A., Bobkoskie, R. L., Tomchick, G. J., Bach, F., and Rakow, A. B., Mercury vapor exposures in dental offices, *J. Amer. Dental Assoc.*, **81**:923, 1970.

437. Joselow, M. M., Goldwater, L. J., Alvarez, A., and Herndon, J., Absorption and excretion of mercury in man XV. Occupational exposures among dentists, *Arch. Environ. Health*, **17**:39, 1968.
438. Lintz, W., Prevention and cure of occupational diseases of the dentists, *J. Amer. Dental Assoc.*, **22**:2071, 1935.
439. Storlazzi, E. D., and Elkins, H. B., The significance of urinary mercury I. Occupational mercury exposure II. Mercury absorption from mercury-bearing dental fillings and antiseptics, *J. Industr. Hyg. Toxicol.*, **23**:459, 1941.
440. Meyer, A., Mercury poisoning, a potential hazard to dental personnel, *Dental Prog.*, **2**:190, 1962.
441. Chandler, H. H., Rupp, N. W., and Paffenbarger, G. C., Poor mercury hygiene from ultrasonic amalgam condensation, *J. Amer. Dental Assoc.*, **82**:533, 1971.
442. Fernstrom, A. I. B., Frykholm, K. O., and Huldt, S., Mercury allergy with eczematous dermatitis due to silver-amalgam fillings, *Brit. Dental J.*, **113**:204, 1962.
443. Jervis, R. E., Debrun, D., LePage, W., and Tiefenbach, B., Mercury residues in Canadian foods, fish, wildlife. National Health Grant Project No. 605-7-510, Department of Chemical Engineering and Applied Chemistry, University of Toronto, 1970.
444. Rosen, E., Argyrolentis, *Amer. J. Ophthal.*, **33**:797, 1950.
445. McCord, C. P., Mercury poisoning in dentists, *Industr. Med. Surg.*, **30**:554, 1961.
446. Ashe, W. F., Largent, E. J., Dutra, F. R., Hubbard, D. M., and Blackstone, M., Behavior of mercury in the animal organism following inhalation, *Industr. Hyg. Occup. Med.*, **7**:19, 1953.
447. Frykholm, K. O., Mercury from dental amalgam: Its toxic and allergic effects and some comments on occupational hygiene, *Acta Odontol. Scand.*, **15(Supp. 22)**:1, 1957.
448. Hardy, H. N., On a case of poisoning by white precipitate, *Brit. Med. J.*, **2**:76, 1876.
449. Stephens, E., Three cases of poisoning by precipitate powder taken in mistake for milk of sulphur, *Brit. Med. J.*, **1**:781, 1876.
450. Sandberg, A. G., Poisoning by white precipitate, *Brit. Med. J.*, **1**:709, 1889.
451. Moore, E. H., Fatal poisoning by white precipitate, *Brit. Med. J.*, **2**:15, 1885.
452. Peters, R. A., Stocken, L. A., and Thompson, R. H. S., British Anti-Lewisite (BAL), *Nature*, **156**:616, 1945.
453. Longcope, W. T., and Luetscher, J. A., Jr., The use of BAL (British Anti-Lewisite) in the treatment of the injurious effects of arsenic, mercury and other metallic poisons, *Ann. Intern. Med.*, **31**:545, 1949.

454. Carleton, A. B., Peters, R. A., Stocken, L. A., Thompson, R. H. S., and Williams, D. I., Clinical uses of 2,3-dimercaptopropanol (BAL), VI. The treatment of complications of arseno-therapy with BAL (British Anti-Lewisite), *J. Clin. Invest.*, **25**:497, 1946.

455. Munoz, J. M., Treatment of acute mercurial poisoning with sodium formaldehyde-sulfoxylate, *Comp. rend. Soc. Biol.*, 120:500, 1935; *Chem. Abstr.*, **30**:762, 1936.

456. Arena, J. M., Treatment of mercury poisoning, *Modern Treatment*, **8**:619, 1971.

457. Stocken, L. A., British Anti-Lewisite as an antidote for acute mercury poisoning, *Biochem. J.*, **41**:358, 1947.

458. Gilman, A., Allen, R. P., Philips, F. S., and St. John, E., Clinical uses of 2,3-dimercaptopropanol (BAL). X. The treatment of acute systemic mercury poisoning in experimental animals with BAL, thiosorbitol and BAL glucoside, *J. Clin. Invest.*, **25**:549, 1946.

459. Longcope, W. T., Luetscher, J. A., Jr., Calkins, E., Grob, D., Bush, S. W., and Eisenberg, H., Clinical uses of 2,3-dimercaptopropanol (BAL). XI. The treatment of acute mercury poisoning by BAL, *J. Clin. Invest.*, **25**:557, 1946.

460. Bivings, L., and Lewis, G., Acrodynia: A new treatment with BAL, *J. Pediatr.*, **32**:63, 1948.

461. Burke, W. J., and Quagliana, J. M., Acute inhalation mercury intoxication, *J. Occup. Med.*, **5**:157, 1963.

462. Deichmann, W. B., and Gerarde, H. W., *Toxicology of Drugs and Chemicals*, New York, Academic, 1969.

463. Sanchez-Sicilia, L., Seto, D. S., Nakamoto, S., and Kolff, W. J., Acute mercurial intoxication treated by hemodialysis, *Ann. Intern. Med.*, **59**:629, 1963.

464. Brown, J. R., and Kulkarni, M. V., A review of the toxicity and metabolism of mercury and its compounds, *Med. Serv. J. Can.*, **23**:786, 1967.

465. Bell, R. F., Gilliland, J. C., and Dunn, W. S., Urinary mercury and lead excretion in a case of mercurialism, *Industr. Health*, **11**:231, 1955.

466. Woodcock, S. M., A case illustrating the effect of calcium disodium versenate (CaNa₂ E.D.T.A.) on chronic mercury poisoning, *Brit. J. Industr. Med.*, **15**:207, 1958.

467. Anon., Mercury picture improves, *TVA News, Weekly News Letter*, July 20, 1972.

468. Bonger, L. H., and Khatta, M. N., *Sand and Gravel Overlay for Control of Mercury in Sediments*, Office of Research and Monitoring, U.S. Environmental Protection Agency, 16080 HVA 01/72, Washington, D.C., 1972.

469. Harlin, C. C., Jr., *Control of Mercury Pollution in Sediments*, Office of

Research and Monitoring, U.S. Environmental Protection Agency, EPA-R2-72-043, Washington, D.C., 1972.

470. Widman, M. U., and Epstein, M. M., *Polymer Film Overlay System for Mercury Contaminated Sludge—Phase I*, Office of Research and Monitoring, U.S. Environmental Protection Agency, 16080 HTZ 05/72, Washington, D.C., 1972.

471. Feick, G., Johnason, E. E., and Yeaple, D. S., *Control of Mercury Contamination in Freshwater Sediments*, Office of Research and Monitoring, U.S. Environmental Protection Agency, EPA-R2-72-077, Washington, D.C., 1972.

472. Tratnyek, J. P., *Waste Wool as a Scavenger for Mercury Pollution in Waters*, Office of Research and Monitoring, U.S. Environmental Protection Agency, 16080 HUB 04/72, Washington, D.C., 1972.

473. Roberts, E. J., and Rowland, S. P., Removal of mercury from aqueous solutions by nitrogen-containing chemically modified cotton, *Environ. Sci. Technol.*, **7**:552, 1973.

474. Suggs, J. D., Petersen, D. H., and Middlebrook, J. B., Jr., *Mercury Pollution Control in Stream and Lake Sediments*, Office of Research and Monitoring, U.S. Environmental Protection Agency, 16080 HTD 03/72, Washington, D.C., 1972.

475. Hansen, R. S., Dredging: Problems and Remedies, *Limnos*, **4**:3, 1971.

476. Jernelöv, A., and Lann, H., Studies in Sweden on the feasibility of some methods for restoration of mercury-contaminated bodies of water, *Environ. Sci. Technol.*, **7**:713, 1973.

477. Anon., An act making appropriations for the construction, repair, and preservation of certain public works on rivers and harbors, and for other repairs, *U.S. Statutes at Large*, **24**:329, 1886 and **26**:453, 1890.

478. Anon., Pollution control: The court scene, *The Winnipeg Tribune*, October 14, 1971.

479. Anon., An act making appropriations for the construction, repair, and preservation of certain public works on rivers and harbors, and for other purposes, Section 13, *U.S. Statutes at Large*, **30**:1152, 1899.

480. Atkeson, T., Speech to ALI-ABA Seminar on Environmental Law, Smithsonian Institution, Washington, D.C., January 28, 1971, Hearings on Refuse Act Permit Program before the Subcommittee on the Environment of the U.S. Senate Committee on Commerce, Ninety-second Congress, 1st Session, Ser. 92-7, 1971.

481. Anon., Public Health Service Act., *U.S. Statutes at Large*, **37**:309, 1912.

482. Anon., Oil Pollution Act, *U.S. Statutes at Large*, **43**:604, 1924.

483. Anon., Water Pollution Control Act, Public Law 80-845, *U.S. Statutes at Large*, **62**:1155, 1948.

484. Anon., Water Pollution Control Act Amendments of 1956, Public Law 84-660, *U.S. Statutes at Large*, **70**:498, 1956.

485. Anon., Fish and Wildlife Coordination Act, Public Law 85-624, *U.S. Statutes at Large*, **72**:563, 1958.

486. Anon., Federal Water Pollution Control Act Amendments of 1961, Public Law 87-88, *U.S. Statutes at Large*, **75**:204,1961.

487. Anon., Water Quality Act of 1965, Public Law 89-234, *U.S. Statutes at Large*, **79**:903, 1965.

488. Massey, D. T., Research need related to institutional-legal aspects of recycling municipal wastewaters on the land. Proceedings of the Workshop on Research Need Related to Recycling Urban Wastewater on Land sponsored by the GLUMORBA Sub-Committee on Land Disposal, Chicago, Illinois, March 19–21, 1974. Proceedings available through the Institute on Land and Water Resources, Pennsylvania State University, University Park, Pa., 16802.

489. Anon., Clean Water Restoration Act of 1966, Public Law 89-753, *U.S. Statutes at Large*, **80**:1246, 1966.

490. Zwerdling, D., And now, mercury, *The New Republic*, **163(5)**:17, August 1, 1970.

491. Rogers, P. G., HEW, Interior should act immediately to halt all dumping of mercury in waters, *Congr. Rec.*, **116(17)**:23558, 1970.

492. Hickel, W. J., Quote in testimony of C. L. Klein at the hearing before the Subcommittee on Energy, Natural Resources, and the Environment of the Committee on Commerce, United States Senate, Ninety-first Congress, 2nd Session on *The Effects of Mercury on Man and the Environment*. Philip A. Hart, Chairman, Part 2, July 30, 1970, Serial 91-73, U.S. Government Printing Office, Washington, D.C., 1970.

493. Stuart, P.C., Refuse Act revived old dog with new tricks, *Christian Science Monitor*, September 24, 1970.

494. Reuss, H., Water pollution and the Refuse Act of 1899: The Corps of Engineers is doing its duty, why not the Department of Justice? *Congr. Rec.*, **116(21)**:28935, 1970.

495. Anon., Reuss gets few answers on mercury pollution cases pending in courts, *Air/Water Pollut. Rep.*, **9(28)**:281, 1971.

496. Anon., Pollution law snub draws blast, *Birmingham News*, July 15, 1970.

497. Corrigan, R., Mercury pollution problem catches federal government unprepared, *Cent. Polit. Res. Natl. J.*, **2(32)**:1695, 1970.

498. Anon., Hickel attacks mercury polluters, *Detroit Free Press*, July 23, 1970.

499. Anon., Justice Department files suits against 10 for mercury pollution, *Air/Water Pollut. Rep.*, **8(30)**:295, 1970.

500. Klein, C. L., Testimony at the hearing before the Subcommittee on Commerce, United States Senate, Ninety-first Congress, 2nd Session on *The Effects of Mercury on Man and the Environment*, Philip A. Hart,

Chairman, Part 2, July 30, 1970, Serial 91-73, U.S. Government Printing Office, Washington, D.C.

501. Anon., Mercury: Blast at industry, *Chem. Eng. News*, **48(32)**:14, 1970.

502. Anon., Known mercury dischargers (analysis positive, as of April 26, 1971). In *Mercury Pollution and Enforcement of the Refuse Act of 1899*, a hearing before a subcommittee of the Committee on Government Operations, House of Representatives, Ninety-Second Congress, 1st Session, Washington, D.C., U.S. Government Printing Office, 1971.

503. Anon., Hickel lists 50 mercury polluters, reports discharges cut 86%, *Air/Water Pollut. Rep.*, **8(38)**:381, 1970.

504. Anon., Hickel hails drop in mercury spills by 86% since July, *Washington Water Line*, **IV:(38)**: 152, 1970.

505. Anon., Olin spending $1.4 million to cut mercury pollution; only 17 states reported "clean," *Air/Water Pollut. Rep.*, **8(50)**:502, 1970.

506. Anon., Georgia-Pacific denies mercury pollution charges, *Clean Air and Water News*, **2(33)**:2, 1970.

507. Newkirk, L., Hickel confirms success in controlling mercury pollution, Georgia-Pacific Corporation News Release, No. 02-0-8, August 6, 1970.

508. Quarles, J. R., Testimony before the subcommittee on the Committee on Government Operations, House of Representatives. In *Mercury Pollution and Enforcement of the Refuse Act of 1899,* Part 2, Ninety-second Congress, 1st Session, Washington, D.C., U.S. Government Printing Office, 1971.

509. Anon., Mercury discharges into U.S. water, *Water Newsl.*, **12(20)**:2, 1970.

510. Anon., Commerce Secretary Stans has released progress reports on mercury and detergents, *Chem. Eng. News*, **48(50)**:25, 1970.

511. Anon., Suit against Olin Corporation adjourned, *Environ. Rep. Curr. Dev.,* **1(17)**:445, 1970.

512. Anon., Oxford paper agrees to end mercury discharge operation at Rumsford plant, *Air/Water Pollut. Rep.*, **8(32)**:322, 1970.

513. Anon., Action varies in federal suits against mercury discharging firms, *Environ. Rep. Curr. Dev.*, **1(20)**:509, 1970.

514. Anon., Mercury curbs in Georgia, *Air and Water News,* **4**:7, November 2, 1970.

515. Reuss, H. S., Letter to Lt. General F. J. Clarke. In *Mercury Pollution and Enforcement of the Refuse Act of 1899*, Part 1, a hearing before a subcommittee of the Committee on Government Operations, House of Representatives, Ninety-second Congress, 1st Session, Washington, D.C., U.S. Government Printing Office, 1971.

516. Anon., Corps of Engineers announces new permit requirements. In *Mercury Pollution and Enforcement of the Refuse Act of 1899*, Part 1,

a hearing before a subcommittee of the Committee on Government Operations, House of Representatives, Ninety-second Congress, 1st Session, Washington, D.C., U.S. Government Printing Office, 1971.

517. Nixon, R. M., Executive Order 11574: Administration of Refuse Act Permit Program, *Fed. Regist.*, **35**:19627, 1970.

518. Reuss, H.S., Letter to R. M. Nixon. In *Mercury Pollution and Enforcement of the Refuse Act of 1899*, Part 1, a hearing before a subcommittee of the Committee on Government Operations, House of Representatives, Ninety-second Congress, 1st Session, Washington, D.C., U.S. Government Printing Office, 1971.

519. Anon., Permits for discharges or deposits into navigable waters, *Fed. Regist.*, **35**:20005, 1970.

520. Schorr, B., Getting tough: U.S. pressure on firms to clean up waterways begins to have impact. Rule requiring dump permit prompts some companies to quit polluting instead, *Wall Street Journal*, November 4, 1971.

521. Quarles, J. R., Jr., Testimony before the subcommittee on the Committee on Government Operations, House of Representatives. In *Mercury Pollution and Enforcement of the Refuse Act of 1899*, Part 2, Ninety-second Congress, 1st Session, Washington, D.C., U.S. Government Printing Office, 1971.

522. Rodgers, W. H., Jr., Industrial water pollution and the Refuse Act: A second chance for water quality, *Univ. Pa. Law Aev.,* **119**:761, 1971.

523. Fri, R. W., *Clean Water: Report to Congress*, U.S. Environmental Protection Agency, Washington, D.C., May 1973.

524. Bowman, J., Testimony before the Subcommittee on Energy, Natural Resources and Environment of the Committee on Commerce, United States Senate, Ninety-first Congress, 2nd Session on *The Effects of Mercury on Man and the Environment*. Philip A. Hart, Chairman, Part 1, May 8, 1970, Serial 91-73, Washington, D.C., U.S. Government Printing Office, 1970.

525. Hill, G., Polluters populate most pollution control boards, *Toronto Daily Enterprise*, December 10, 1970.

526. Anon., Lax pollution laws attacked, *Fargo Forum,* May 19, 1971.

527. Gushee, D. E., Clean Water: What is it? How will we achieve it? *Chem. Tech.*, **3**:334, 1973.

528. Anon., Worst polluters buy environmental ads, study claims, *Minneapolis Tribune*, November 5, 1971.

529. Anon., Water Strategy Paper, Statement of policy for implementing certain requirements of the 1972 Federal Water Pollution Control Act Amendments, U.S. Environmental Protection Agency, Washington, D.C., April 30, 1973.

530. Anon., Showdown shapes up over pollution laws, *Chem. Eng. News*, **52(4)**:17, 1974.

531. Anon., Fourth annual report of the Council on Environmental Quality. *Environmental Quality*, U.S. Government Printing Office, Washington, D.C., 1973.

532. Anon., *Action for environmental quality standards and enforcement for air and water pollution control.* U.S. Environmental Protection Agency, Washington, D.C. 20460, Office of Public Affairs, S/N 5500-00087, March 1973.

533. Anon., EPA proposed toxic pollutant effluent standards, *Fed. Regist.*, **38**:35388, 1973.

534. Schafer, C. J., and Strier, M. P., Current regulations and enforcement experience by Environmental Protection Agency. In P. A. Krenkel, Ed., *Heavy Metals in the Aquatic Environment,* Pergamon, New York, 1975.

535. Glenn, M., Permit Programs Division, Office of Water Enforcement, EPA. Discussion of paper entitled Current regulations and enforcement experience by EPA, International Conference on Heavy Metals in the Aquatic Environment, Vanderbilt University, December 4–7, 1973.

536. Anon., Guidelines for Developing or Revising Water Quality Standards Under the Federal Water Pollution Control Act Amendments of 1972. U.S. Environmental Protection Agency, Water Planning Division, Planning and Standards Branch, Washington, D.C. 20460, January 1973, Amended April 1973.

537. Harada, M., Minamata Disease: A medical report. In W. E. Smith and A. E. Smith, *Minamata,* New York, Holt, Rinehart and Winston, 1975.

INDEX

Acetaldehyde, 23, 89
n-Acetyl-d, 1, penicillamine, *see* British
  anti-Lewisite
Acrodynia:
  definition, 164, 197–198
  identification, 199
  nonprescription medications, 200–201
  symptoms, 200
  treatment, 201
Agano River, Japan, 22
Agricultural chemicals, 19. *See also*
  Alkylmercury fungicides
Alamogordo, New Mexico, 1, 41–42,
  44–46
Alchemy, 9–10
Alkylmercury fungicides:
  compounds:
    Agrosan GN^R, 36
    alkyloxyalkylmercury, 31
    Betoxin F^R, 30
    Cerosan^R, 46
    ethylmercury, 35
    Granosan^R, 35
    Panogen^R, 30, 36, 45
    Panogen Metrox^R, 35
  international shipment, 46
  legislation:
    Canada, 32
    England, 31
    Sweden, 31
    United States, *see* United States,
      legislation for pollution control
  manufacture:
    exposure, 32

grain testing, 33
grain treatment, 33
storage, 33
poisoning:
  birds, 29–30
  epidemics, 35–36, 46, 87
  individuals, 33–36, 41
Almadén, 120–121, 124, 126
Amalgamation, 7
Ancient god, 5
Ancient Greece, 4
Ancient historical accounts, 7
Antiseptics, 4, 13
  cold sterilization, 208–209
  compounds:
    Metaphen^R, 111, 209
    Mercarbolite^R, 209
    Mercresin^R, 209
    Merthiolate^R, 208
    mercuric chloride, 208–209
    mercuric salts, 208
    Mercurochrome^R, 111, 208
    organomercurials, 208
  Council of Pharmacy and Chemistry,
    AMA, 209
  effect on microorganisms, 208–209
  legislation, 210
  nonprescription medications,
    208
  on fabrics, 210
  Koch, Robert, 208
Argentum Vivum, 6
Ariake-cho, Japan, 16, 23
Aristotle, 6–7